Einstein's Third Mistake

Space, Time and Gravity

Alex M Murphy

 The Old Grouse Press, North Yorkshire, England

Einstein's Third Mistake

ISBN 978-1-4709-1425-7
First published, October 2011.

'The speed of light cannot be increased
but it can be supplemented'
(pages 47-55)

Einstein's Third Mistake

Summary

- Einstein claimed that time can run fast or slow and that space can expand or contract.
- There is no evidence from observation or experiment to support space/time (S/T) dilation.
- S/T dilation was needed for theoretical reasons, to resolve the Galileo/Maxwell dilemma.
- This dilemma can be resolved by showing that the speed of light can be supplemented.
- Einstein claimed that gravity was not a force but the result of the bending of space.
- If space cannot bend then Einstein's explanation of gravity is wrong.
- Gravity is the result of the unevenness in different areas of the fabric of space.
- This unevenness is partly the result of the expansion of the universe.
- The expansion is inhibited locally by matter and produces a gradient of traction.
- All matter is in motion as a result of its movement in space and/or its atomic structure.
- Energy from either type of motion is converted to a force by the traction of space.
- This is the force of gravity.

Einstein's Third Mistake

CONTENTS

Explaining Gravity
The Particle Model of Forces
Einstein's Explanation
The Bending of Space Experiment
The Three Mistakes

The Speed of Light
Time Dilation
A Preliminary Objection
Clocks and the Working Environment
What is "Time"?
The Bending of Space
The Meaning of 'speed'

.

The Galileo/Maxwell Dilemma
Frames, Observation and Events
Science and Falsifiability

Einstein's Third Mistake

CONTENTS

(continued)

Page

Einstein's Third Mistake

Preface

In a 1916 memorial note for Ernst Mach, a physicist and philosopher he respected, Einstein wrote:

"Concepts that have proven useful in ordering things easily achieve such an authority over us that we forget their earthly origins and accept them as unalterable givens. Thus they come to be stamped as "necessities of thought," "a priori givens," etc. The path of scientific advance is often made impassable for a long time through such errors. For that reason, it is by no means an idle game if we become practiced in analysing the long commonplace concepts ... By this means, their all-too-great authority will be broken. They will be removed if they cannot be properly legitimated ..."

Ernst Mach, *Physikalische Zeitschrift, 17 (1916) p.102.*
Quoted in: Minnesota Studies in the Philosophy of Science, Volume 5.

Einstein's Third Mistake

Chapter 1

Explaining Gravity

One of the most important developments in twentieth century science was Einstein's revolutionary explanation of how gravity works. We experience gravity in our daily lives and we are used to thinking of it as a 'force'. Gravity causes objects such as apples and children to fall to the ground and holds planets in their orbits around the Sun. We experience gravity when we try to lift a weight and feel the downward pull against us. We can feel the force of gravity just as we feel the force of a magnet pulling on a piece of metal or another person pulling on a rope we hold. Our assumption that gravity is a force comes from this everyday experience of the world.

This commonsense understanding of gravity as a force was rejected by Einstein and replaced with his own revolutionary view. According to Einstein, gravity was bent space affecting the behaviour of suns, planets, apples and children as they moved through space. All matter follows the path which bent space provides much like a train follows the curve in the railway track. Apples which fall from the tree are not pulled or pushed by a force but are following a pathway of bent space from the branch to the centre of the Earth. This bending of space, Einstein argued, is

1

caused by all forms of matter from a pin to a moon or planet. Massive objects such as planets or suns cause massive bending of space. When an apple falls from the tree it is following a path of bent space caused by presence of a massive object such as the planet Earth. This explanation by Einstein of how gravity works is mistaken and explaining how it is mistaken is the purpose of this book.

It is not just massive objects which bend or distort space. Any material object, from an apple to a planet or a pin, has this space-distorting effect and creates a gravitational influence on other matter. The gravitation of the Moon affects objects here on Earth but objects here on Earth also have a gravitational influence on the Moon. Since the Moon is so much larger than an apple here on Earth the influence of the apple on the Moon is negligible but the Moon and the apple are both affected by the gravitational pull of the other.

When Einstein's theory of gravity and bent space was 'verified' in 1918 by the observations of Sir Arthur Eddington on the island of Principe, the importance of the theory was recognised by the scientific community across the world. "The Times" described it as a "Revolution in Science" and a "New Theory of the Universe". Since that time his theory has become the established and establishment model of how gravity is to be explained.

The Times, November 7, 1919

Before looking in detail at Einstein's theory, it is interesting to examine some aspects of gravity which we can observe in the everyday world but of which we are not usually aware.

Gravity is not uniform across the whole planet. Various factors influence the force of gravity and there are slight deviations in the strength and direction of gravity across the surface of the planet. The nearer an object is to the centre of a planet the stronger gravity is. Gravity is weaker near the equator where the equatorial bulge in the planet's surface causes objects to be farther away from the centre of the planet. Objects at the poles experience a slightly stronger gravitational pull than objects on the equatorial bulge because they are nearer to the centre of the planet. The 'force' of gravity also decreases with altitude. The distance from the top of a mountain to the centre of the planet means that gravity is weaker at the top of a mountain than in a valley. Gravity causes bulges in the Earth's surface as well as in sea levels. It is a major factor in rock formations and affects the way pendulum clocks behave.

One of the more obvious examples of gravity which we can observe is the effect of the Moon's gravity on the seas. High tides are caused by the gravity of the Moon pulling the water upwards while the Earth's gravity is pulling it downwards. The Moon's gravity wins this struggle and causes the rise of the water which we call tides. On the opposite side of the Earth, simultaneously, there is a second high tide caused by the inertia of the ocean water on the Earth's surface. As the planet and the planet surface below the water is being pulled toward the Moon, the ocean water remains left behind as a result of this inertia of the water. In other words, the water 'sticks' to the ocean floor and remains behind as the planet is pulled towards the Moon. This creates a secondary high tide on the side of the Earth opposite the primary high tide caused by the direct pull of the Moon on the sea.

The situation is made more complicated because of the influence of the gravity of the Sun which sometimes works with and sometimes against the gravitational pull of the Moon. The Moon is approximately 250,000 miles from the Earth and exerts a greater influence on the tides than the Sun, which is over 90 million miles from the Earth. The Sun's gravity is some 180 times stronger than the Moon's but the Moon, because of the Moon's proximity to the Earth. is responsible for 56% of the Earth's tidal energy while the Sun is responsible for only 44%. This interaction of the gravitational effects of Sun, Moon and Earth give us spring and neap tides which are about 20% higher or lower than the average tides.

In addition to sea tides, there are land tides. The Moon's gravity affects the shape of the Earth and pulls the crust up into lumps. An interesting and unusual effect of the Moon's gravity was the problems it caused scientists with the experiments on particle acceleration at CERN, which sits on the border between France and Switzerland near Geneva. Scientists had assumed that there was something wrong with their experimental equipment and this was causing unusual and unexpected results when they fired sub-atomic particles round the 17-mile underground ring of the accelerator. They later discovered that the problems were the result of the pull of the Moon's gravity.

The Moon's gravity did not directly affect the particles, electrons and positrons, as they were fired round the accelerator but it did distort the ground in which the tunnel is embedded, changing the tunnel's 17-mile circumference by about one millimetre. This change in the accelerator's dimensions caused tiny fluctuations in the electron and positron beams. This minute change in the structure of the tunnel itself, caused by the gravitational tidal pull of the Moon on the Earth's surface, was the explanation of the unexpected results *(see also Addendum, p.116)*.

These are curious examples of gravity at work in the everyday world which supplement our more humdrum experience of gravity such as hauling heavy loads uphill and lifting weights. If you throw a ball into the air you feel the force of the ball as it falls and you catch it. Whether the ball is thrown by someone or falls under the influence of gravity, both provide us with an experience of force. Our everyday understanding of gravity is that it is a force of this sort.

The Particle Model of Forces

The established scientific view is that there are four main natural forces. These are the *strong* and *weak nuclear forces*, the *electromagnetic force* and *gravity*. Einstein, however, believed that gravity was not a force like the other three. He argued that gravity

6

was not a force at all but was the consequence of the bending of space. To appreciate Einstein's revolutionary explanation of gravity we need to have some understanding of the traditional scientific model, or models, of a force.

Scientists have established various 'models' of how the main physical forces work, how energy is exchanged between object A and B. There is the particle model and the wave model. The easiest way to understand the particle model of a force is the 'billiard ball' analogy, in which the energy from the player's arm is transferred to the white ball and then on to the red. There is an alternative to this particle model of force, the wave model, which is not quite so easy to visualise. We shall come to this wave model later but start by describing the particle model

The conventional model of particle exchange can be illustrated with two examples of attractive and repulsive forces. In the illustrations below, the particles which carry forces are represented by a heavy ball and a boomerang. These are the force carriers. One boy is the source of the energy, the other boy is the recipient and the ball is the medium of transfer for the energy.

In Diagram 1 there are two boys on skateboards. They are on skateboards because particles are always in motion and this is represented by the moving skateboards. They are facing each other and travelling in direction E.

Diagram 1
Repulsive Force

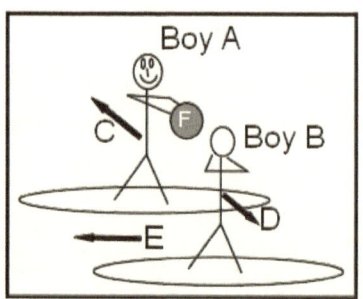

Boy A throws a heavy ball, F, to Boy B. The effort needed to throw the ball forces Boy A to recoil, in the direction of arrow C, and move away from Boy B. When Boy B receives the ball, he recoils with the force of receiving the ball and moves, in the direction of arrow D, away from Boy A.

The particle, in this case the heavy ball, is the carrier of energy from Boy A to Boy B and is the intermediary for the exchange of energy. This is a simplified example of a repulsive force, the exchange of force by a particle causing repulsion between the sender and receiver.

Diagram 2
Attractive Force

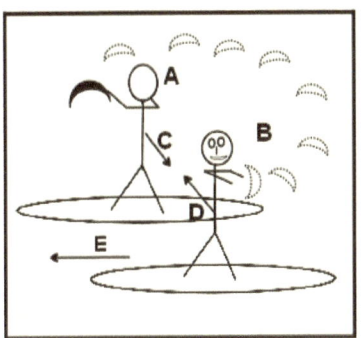

In Diagram 2 we also have two boys on skate-boards. They are facing away from each other and travelling in the same direction (E). Boy A throws a heavy boomerang. The effort of throwing forces him backwards towards Boy B, indicated by the arrow C in the diagram. The boomerang behaves as boomer-angs do and curves round to be caught by Boy B. The force of catching the boomerang forces Boy B back-wards towards Boy A, indicated by the arrow D. This is an illustration of how an attractive force might work.

In reality, the particles which carry the forces are not heavy balls or boomerangs but sub-atomic parti-cles. Whether forces are best explained in terms of particles or waves is a question we shall examine in detail later. What is important at this point is that Ein-stein claimed that gravity was not a force like the other three basic forces. Gravity, he argued, was not a force and did not require either waves or particles for its operation. It had to be explained in another, very radical way.

Some scientists have argued that there is a particle called the graviton, which acts as the carrier of the force of gravity but despite extensive research it has never been found and no evidence exists to show that such a particle really exists. So we will leave this notional particle as a remote theoretical possibility which has never been established and concentrate on what Einstein proposed as the explanation of gravity.

Einstein's Explanation

Gravity, according to Einstein, is not a force of any sort and so any model of a force, whether it uses waves or particles, is not relevant. Gravity, he argued, is the result of the bending of space caused by the presence of matter. Any matter, a pin, an apple or a moon, can cause the bending of space but this bending is more noticeable when massive objects such as galaxies, suns and planets are involved. Matter which moves through 'bent space' follows the distortion of space and behaves as though it is being attracted by some unseen force. This bending of space is illustrated below.

Diagram 3
Bending of Space

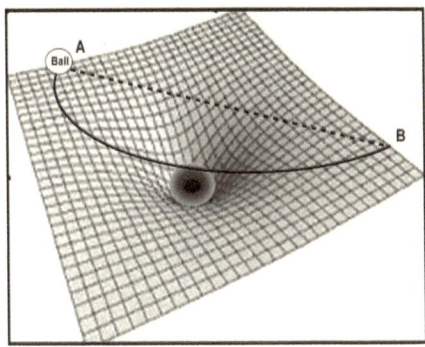

In this illustration we have a rubber mat in the centre of which a heavy object is placed. The weight of the object causes the mat to distort. The distortion forms a dip in the mat with the heavy object at the centre. A ball is launched from point A towards B. In normal circumstances, we would expect the ball to travel in a straight line from A to B as shown by the dotted line. However, because of the indentation in the rubber mat, caused by the presence of the large object, the ball follows the indentation in the rubber sheet and follows a curved path towards point B. This is shown by the continuous curved line which dips towards the object in the centre of the mat.

The ball is not attracted to the centre by any force. It simply follows the curve in the surface across which it travels. It only appears to be attracted to the centre of the mat. This was Einstein's revolutionary explanation of the 'force' of gravity. No force is involved, simply a distortion of space by a massive object. If space is distorted by a massive object such as the Sun then anything, including light, travelling across this space would be affected and would follow the path created by the distortion of space.

The diagram amounts to no more than an illustration, a picture to help people visualise what 'bent space' might mean. It is not an explanation of what bends space or what makes the ball follow that path. Why does the ball follow the bend in space? Why does it not travel directly from A to B? In the real world the ball would follow the bend in the mat/space because it would be pulled down by gravity. Since we are trying to explain how gravity works, this is not an explanation and is merely a visualisation. The diagram is a useful illustration or visualisation of Einstein's explanation of gravity but it raises and does not answer two key questions.

What is the mechanism which causes space to bend? What is the mechanism which makes the ball follow the curve of bent space rather than ignore the bent mat and travel in a direct line from A to B?

The Bending of Space Experiment

Einstein's notion of bent space started as a hypothesis which needed to be confirmed by experiment or observation before it could be accepted. It was proposed that the movement of light through bent space could be identified at the time of an eclipse when light passed near the Sun. The Sun would be the massive object which caused space to be distorted and the Moon would eclipse the Sun. This would allow observers to see light from distant stars, hidden behind the Sun, as the light bent round the Sun and followed the path of distorted space.

In 1919, the Royal Astronomical Society arranged an expedition to the island of Príncipe off the west coast of Africa to observe a total solar eclipse and prove Einstein's bending of space theory. The story of the Principe Island experiment had so many twists and turns, so many problems and setbacks, it could have been a thriller story written by John Buchan or Conan Doyle.

The expedition was led by Sir Arthur Eddington, one of the most distinguished British scientists of his day. During the First World War, and at a time the nation was at war with Germany, Eddington's pacifist views were not popular. He had almost ended up in prison because of his beliefs. But these same beliefs had allowed him to stay in touch with German scientists. He did not reject the idea of working with German scientists as many of his British colleagues did and he looked at Einstein's theory with an open mind. He strongly supported Einstein's views on the bending of space and the curved path of light as it travelled through space. He was determined to try to prove Einstein's theory after the war ended.

The total eclipse of the Sun by the Moon on the 29th May 1919 gave Eddington the ideal opportunity to test Einstein's theory for the first time. Totality, when the Sun was totally obscured by the Moon, lasted for almost seven minutes, an unusually long time for an eclipse. A bright cluster of stars, the Hyades, were behind the Sun during the eclipse. The Moon obscured the Sun and shielded the stars, allowing the observers to view the light from the stars as it passed by the Sun and followed the path of bent space.

On the day of the experiment, clouds obscured the skies and it had rained every day for over two weeks before the eclipse. As the eclipse began, the sun was obscured by clouds. For 400 seconds, the eclipse was hidden from view. Then, with only ten seconds of the eclipse remaining, the skies cleared and Eddington was able to take just one photograph. He compared his eclipse photo with images he had taken in England and announced that the sun had caused a deflection in the path of the light of roughly 1.61 seconds of arc, a result that was in agreement with Einstein's prediction. The light from the stars had bent round the sun and followed a curved path. This curved path made it appear that the stars had shifted their position. Of course, they had not changed their positions it was simply that light from the stars was being bent and made it appear that their position had shifted.

Diagram 4

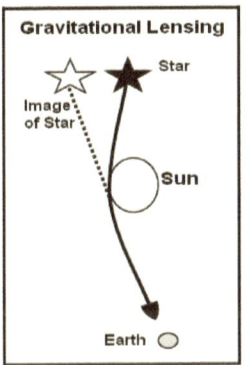

This phenomenon, known now as Gravitational Lensing, is used extensively today to allow astronomers to collect data about distant astronomical objects by observing how the path of light from them is distorted as it travels past massive objects.

In later years, some scientists questioned Eddington's margin of error, arguing that his equipment for the experiment was not accurate enough to allow such an exact conclusion. Some even argued that Eddington believed so strongly in Einstein's theory and wanted so much to prove that it was true that, subconsciously, he may have minimised his errors to get the answer he needed. However, the observations on Principe were accepted and showed that the light from the stars appeared to shift from the positions as measured by Eddington in England. The results of the Principe observations provided the first experimental confirmation of Einstein's theory that space was bent by the presence of a massive object.

The observations, and their confirmation of Einstein's theory, were regarded at the time as revolutionary and made the headlines in newspapers around the world. An example of a report from the New York Times in the USA is below and illustrates the emphasis on an 'epoch' or 'step' change in scientific thinking.

New York Times November 9, 1919

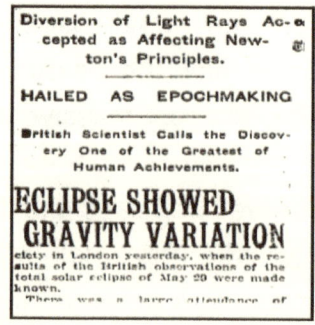

Diversion of Light Rays Accepted as Affecting Newton's Principles.

HAILED AS EPOCHMAKING

British Scientist Calls the Discovery One of the Greatest of Human Achievements.

ECLIPSE SHOWED GRAVITY VARIATION

ciety in London yesterday, when the results of the British observations of the total solar eclipse of May 29 were made known.

There was a large attendance of

The Principe observations provided the support Einstein's theory needed and his theory is still considered to be valid and one of the most radical and valuable contributions to twentieth century science.

The Three Mistakes

The results of the expedition to Principe were of enormous importance but the mechanism of bent space which Einstein proposed to explain these results was wrong. There are three crucial errors in Einstein's explanation of gravity as 'bent space'. These three errors are outlined below.

First, he wrongly believed that space and time could dilate, that they could expand or contract. It is this dilation which explains the bending of space. Space and time, however, are not entities and these characteristics belong to entities. Time is not an entity, it is a special type of measurement and we measure time using 'clocks', which are time-measuring tools. We also make use of a space-measuring tool, a ruler, to measure 'space'. 'Time' and 'space' are words we use to describe these special types of measurements not to describe entities.

In the mid nineteen-fifties, a British philosopher, Gilbert Ryle, published a book called 'The Concept of Mind'. In this book Ryle identifies what he called a 'category error'. A category error happens when we talk about objects, concepts, events or processes and assume that they are all of the same logical type. The mistake is illustrated by Ryle with the example of a prospective student visiting a university. The student is shown the library, the biology labs, the lecture rooms, the sports arena, but at the end of his tour he asks, 'But where is the University?' In asking that question he is working on the assumption that 'the university' is a unique place or building like 'the library' or 'the tennis court'. According to Ryle, the mistake made by the student is a failure to realize that 'university' and 'library' are terms that belong to different logical categories.

Category mistakes are often a source of humour which exploits such confusions. A well-known example of category mistakes comes from Marx, not Karl, but that other eminent philosopher, Groucho.

'You can leave in a taxi. If you can't get a taxi, you can leave in a huff. If that's too soon, you can leave in a minute and a huff.' (Groucho Marx, from 'Duck Soup')

'Taxi' is a concept which belong to one category of concepts while 'huff' belongs to another category and a 'minute and a huff' to yet another, "Groucho-esque", category. Deliberately confusing and muddling these categories provides the humour. Einstein's theory of gravity makes just such conceptual confusions in the use of the concepts of space and time. It ascribes to space and time qualities which they do not and cannot have. Space cannot be bent and time cannot run slow or fast. A time measurement is not a measurement of **something** called time. Nor is a space measurement a measurement of **something** called space. Time and space are words in a special category. They are words used to describe measurements and the information provided by these measurements helps us to manage the world in which we live.

Second, the hypothesis that time and space dilate is not based on experiment and observation. The Principe observations showed that light rays were affected "in some way" as they passed through space and near to the sun but the observations are open to several interpretations and do not support a 'bent space' model exclusively. The bent space hypothesis was introduced to reconcile the conflicting views of Galileo and James Clerk Maxwell. Resolving this philosophical problem was Einstein's reason for introducing space dilation. Space/time dilation is not based on experiment or observation so this raises the question of whether Einstein's theory of gravity is a scientific theory or a philosophical prejudice. There is another way to reconcile the views of Galileo and

Maxwell. This can be done if we clarify what we mean by the 'speed of light' and show that the speed of light cannot be increased, as Maxwell showed, but it can be 'supplemented' as Galileo required. How the speed of light can be supplemented is explained below.

Finally, to provide a mechanism to explain the Principe results we need to take into account space in the sense not just of distance between objects but also the void or emptiness which exists between planets, moons suns and galaxies, the so-called 'fabric of space'. To do this we need to examine new options such as the expansion of the universe and Quantum Theory (QT). Einstein never accepted that QT was a valid scientific theory and the expansion of the universe had not been firmly established when he constructed his theory of gravity.

Chapter 2

The Speed of Light

The speed of light is a key element in Einstein's model of how gravity works and his hypothesis that time and space dilate and we need to understand some of the special qualities which light has. For many years, scientists believed that light moved from A to B instantaneously. Galileo did not agree and set up an experiment with lanterns on hilltops one mile apart. The lanterns were flashed and the time of the flash noted by an observer on the distant hill. Unfortunately, the speed of light is so great that this technique could not detect the time it took for light to travel the one mile distance. What was needed was a much longer distance for the light to travel and a measuring technique which could measure the time taken to travel that distance.

It was only in the 17th century that techniques were available to measure the actual speed of light. The Danish astronomer Ole Roemer was observing Jupiter's moon Io. To his surprise, he noticed that the moon did not always appear where it was supposed to be and where he expected to see it. At various times of the year it seemed to be ahead of the expected schedule and at other times behind schedule. It was a simple but important step for him to work out that at different times the moon was closer to Earth and at other times farther away and so it took the light different times to travel the different distances. He deduced that the moon appeared to be in different places because the light was taking different times to travel to Earth.

Armed with this information, Roemer deduced that light took a finite time to travel from A to B and he was able to calculate the speed of light with some accuracy. With today's more advanced technology we are able to estimate the speed of light with greater accuracy. Astronauts left a mirror on the Moon and we are now able to fire a laser beam at the Moon and record the time of its return. The time taken for this round trip to the Moon is about two and a half seconds which gives light a speed of approximately 186,000 miles per second (mps).

Laser light

In the 1960s, a new option became available which made the use of light even more reliable and valuable. This option was laser light. Laser light has many applications in our everyday life, from writing information on computer discs to making photocopies and providing Christmas decorations over the High Street. There are many different types of laser light but of special interest is laser light which can be very narrowly focussed and which can travel large distance without spreading out so far that it becomes useless. The beam of light from a torch may start out one inch in diameter but after travelling ten feet it may be twenty inches in diameter. Laser light can be focussed much more tightly.

While laser light also spreads, it spreads out less and over much larger distances. A narrow beam of laser light fired from the Earth to the Moon starts only millimetres in diameter but when it is reflected back to Earth by mirrors on the Moon it will spread out to a diameter of over 9 miles on its return. When a measurement is made using laser light fired at the Moon, the beam recorded on its return will be made up of some 30 million photons and only a few of these will be recorded when they hit receivers on Earth. It is easy to think of a laser as firing a 'light bullet', a bullet which travels at 186,000 mps and which can be monitored and recorded like a solid object, like a bullet. Laser light is not a bullet-like entity. Whether laser light or ordinary light is used the same rules apply.

Unique Speed of Light

Light is unusual in that the speed of light has the unique characteristic of being constant. Unlike other objects in motion, such as bullets or rockets, the speed of light cannot be increased above a fixed limit of 186,000 mps. This speed of light in a vacuum is usually denoted by the letter 'C' because it is a universal constant and cannot be increased. It is an absolute maximum, although light does move more slowly if it travels through other mediums. In water, for example, light travels at just over 140,000 mps.

The unique nature of light and the unusual characteristics of its speed can be illustrated in a simple experiment. A car travelling at 50 mph switches on a lamp. A second lamp, which is stationary at the roadside, is switched on and the two beams of light shine in parallel. The speed of the light from the car and speed of light from stationary source are measured and are identical. The speed of light is 'constant'. The speed of the car does not affect the speed of the beam of light fired from the car. This is quite unlike the situation where two bullets are fired from two guns. One gun is on board a car travelling at 50 mph and the other is fired from a roadside location which is stationary in relation to the car. The speed of both bullets is measured and we find that the bullet from the stationary source travels at 300 mph, while the bullet from the speeding car has a speed of 350 mph. The speed at which the car travels, 50 mph, can be added to the speed of the bullet as it leaves the muzzle of the gun (300 mph). The speed of the bullet is increased by the speed of the car from which it is fired but the speed of light is constant and the speed of the car cannot be added to the speed of the light. This appears to make light a unique and invaluable tool for measuring.

The equations which Maxwell developed to explain electromagnetic radiation, including light, does not allow the speed of light to be increased. It is part of the nature of light that it has this characteristic, the speed of 'C', 186,000 mps, which cannot be increased. We shall see later that this proposition may have several meanings some of which are misunderstood or cannot be taken at face value.

Maxwell's hypothesis that the speed of light is constant and cannot be increased creates a dilemma when compared with the views of Galileo. Galileo believed that speeds could be cumulative. For example, a boy fires an arrow from a moving ship and for an observer on the shore the speed of the arrow is the speed as it leaves the bow added to the speed of the ship on which the boy stands. The two speeds are cumulative. Since the speed of light is constant and cannot be increased then if a light is shone from a ship the speed of the ship cannot be added to the speed of the light. This is the dilemma which Einstein tried to resolve and his solution required that both space and time can expand and contract.

Time Dilation

At various times, scientists and philosophers have believed that time was 'objective', that it was 'out there', and that we could make objective measurements of time. They believed that 'time' was unchanging, that some independent measuring rod of time existed and could be used to check local time measurement. Einstein, on the other hand, thought that time was not objective and it was not unchanging. 'Time', he argued, was relative and depended on the environment in which it was measured. In one environment, or frame as he called it, time would run more slowly than in another. An example of a frame of referenced is the observations of a passenger on a train. From his frame he sees himself as stationary. His coffee sits on the table and does not move or spill. He puts his newspaper down on the table and it remains at rest on the table. The passenger on the train sees himself as at rest but he sees the fields and stations outside the carriage in motion. The traveller waiting on the station platform provides the second frame of reference. He sees the train and the passenger in motion but sees himself at rest. He sees the passenger in the carriage, his coffee and his newspaper flash past. We have two frames of reference with observations from observers in both frames and their observations do not agree. There is no absolute frame of reference which can tell us which observer is in motion and which is at rest. Frames of reference are important and we shall return to discuss them in detail later.

There are no experiments or observations to support his theory that time and space dilate. To support his view Einstein relied on a thought experiment in which the speed of light is a key element. This thought experiment was not intended merely as an example or helpful illustration. It was his core argument used both to explain and support his hypothesis.

For that reason it takes centre stage. A simple example of a light clock and the observations from two frames of reference is as follows.

In Einstein's classic thought experiment two light clocks are in motion relative to each other. An observer with clock 1 believes that he and his clock are at rest and sees the second clock in motion. The observer with clock 2 believes he is at rest and that clock 1 is in motion. There is no way we can establish which clock is at rest and which is in motion. There is no absolute standard to judge when an object is at rest. There is no single, absolute frame of reference from which one clock can be judged to be at rest and the other in motion. What we know is that the clocks are in motion relative to each other and observations from the frame of reference of either clock differ when compared.

Diagram 5

Light Clock 1 at Rest

A beam of light is fired from point A to B and when it reaches B it is reflected back to A. When it arrives at A the clock 'pings' or 'chimes' so giving us a measure of time. Since we know that the speed of light is 186,000 mps and it is constant then if we know the distance from A to B we have a very reliable and precise way of measuring time. This is a light clock. Next we arrange to have two identical light clocks and ensure that when they are stationary they are synchronised and 'ping' or chime together so ensuring they are measuring time in the same way. When both clocks are rest they are observed in the same frame of reference which means the observer sees them behave identically.

We then send one clock off on a carriage, any carriage, a sailing ship, a railway train, a spaceship or another planet, will be fine. When the light clock is in motion it still 'pings' exactly as the stationary clock and they both keep the same time. In Diagram 6, an observer from his frame of reference, records what is happening with Clock 2 which is in motion and is in a different frame of reference. His observations of the clock in motion are quite different from his observation of the stationary clock in his own frame of reference. He sees the light in Clock 2 emitted from C travelling to D but, because Clock 2 is in motion as the beam travels from C to D, the beam of light must cover a greater distance. It travels from C not to D but to D1 and back not to C but to C1. Because the clock is in motion this extra distance has to be taken into account in any measurement. The start and end points of the beam have moved because the clock is in motion and the light beam travels over a greater distance, denoted by the letter E. The extra distance travelled by the light beam is the distance between C and C1, or the diagonal paths C to D1 and D1 to C1. Whichever way it is measured, the distance is greater than A to B and B to A in Diagram 5.

This introduces an idea which many find difficult to grasp. We are used to units of weight being units of different items. We have a pound of apples and a pound of pears and the unit of the 'pound' is the same while what is being weighed is different.

Diagram 6
Light Clock in Motion

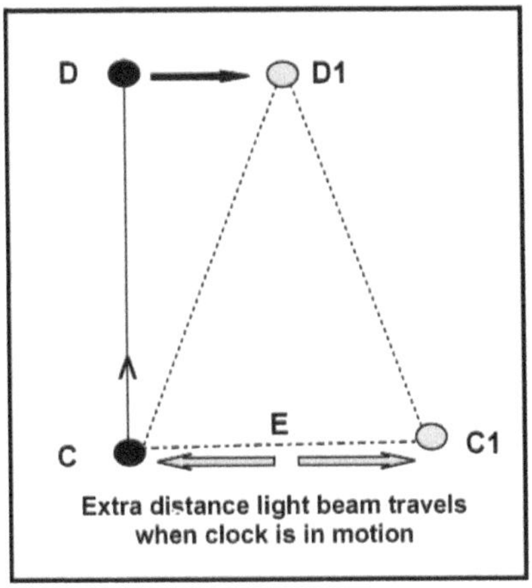

Extra distance light beam travels
when clock is in motion

We now have to accept, according to Einstein, that units of time can also measure different 'things', slow time and fast time. We can have ten seconds of slow time and ten seconds of fast time. A second is a unit of measurement of time but it can measure different varieties of time.

The moving clock is running at the same speed as the stationary clock because we tested both before clock 2 was set in motion. The light beam in Diagram 6 travels in the delta shape and travels further than the beam in Diagram 5. As the two clocks 'ping' together, they are recording the same unit of time and this can only be explained if the clock on the moving ship is recording 'slow time'. This may seem counter-intuitive. If runners were involved and one had to run a longer distance than the other we would expect one runner to run faster to cover the greater distance. In this case however, the 'runner' is the beam of light and this has a constant speed and so cannot 'run faster'. The alternative is that the two units of time are different – they are units of different time. The unit employed by the clock in motion is a unit of time which allows the runner, the light beam, to cover the greater distance. If time runs slower this allows the runner, the light beam, to cover a greater distance in the same unit of time and at the same speed, the speed of light.

Einstein's conclusion was that time can run at different speeds. We are entitled, Einstein argued, to assume that time in the moving clock is running slower than time in the stationary clock to just the amount needed to reconcile the differences in the distances travelled by the light beams of the two clocks. The faster the motion of one clock, the greater the distance the light beam has to travel, and the slower time must be running, when measured by an observer who is stationary relative to the clock which is in motion.

Speed is a measurement which uses both time and distance and the time dilation argument could equally be made into a space dilation argument. In stead of using slow time in the measurement, Einstein's hypothesis would support the view that time remains 'normal' time but the space over which the light beams travels is contracted. The contraction of space means that the extra distance the light beam travels in a light-clock in motion is compensated for by the space over which the beam travels being shortened or contracted space. Time can be dilated to account for the extra distance travelled by a light beam in a clock in motion but the same conclusion is reached if we argue that space over which the beam travels is contracted space. It is possible to claim that time runs slow or that space contracts or some combination of both.

This dilation of time and/or space means that the speed of light can remain constant and the extra distance travelled by a beam of light in a light-clock in motion can be compensated for by the dilation of time or space or both. This hypothesis of space/time dilation which Einstein proposed is wrong and we can now look at why it is wrong.

A Preliminary Objection

The 'traditional argument used to support the claim that time can run at different speeds relies on the use of two light clocks, one at rest and the other in motion relative to the first. We need to ask, then, about other, non-light, clocks and whether they too support this argument.

Diagram 7
Egg-timer Clocks

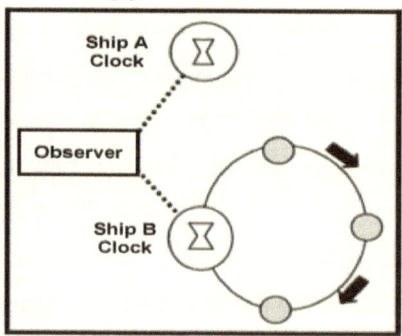

If we use another type of clock, a Swiss watch or an egg-timer, we can find no evidence of time running slow. An egg-timer or Swiss watch thought experiment appears to show that time runs at the same speed whether a non-light clock is stationary or in motion.

Two clocks are placed on ships A and B. The clocks have been synchronised and 'ping' or chime in unison. The clock on B starts its trip while A remains stationary in relation to the observer. Ship B completes its round trip and, as this has been calculated and takes one minute, the clock on board pings. The clock on ship A also pings because it has been set to chime after one minute and so both clocks chime at the same time. The clock in motion has recorded time in exactly the same way as the stationary clock and its motion has made no difference to the operation of the clock on board or the time it is recording. Both clocks ping at the same time and both record the same time although one clock is stationary and the other in motion. This preliminary objection establishes that if we use non-light clocks these provide no evidence from their behaviour that they are measuring two types of time, one slower than the other.

Even an egg-timer, however, would be affected if the working environment was different and acceleration not motion was involved. For example, as a spaceship accelerates the passengers are pushed against the seat. If an egg timer was used on a spaceship on take-off the fall of the sand would be affected by the acceleration of the ship. The acceleration would force the sand more quickly into the lower chamber. The force of the acceleration would be transferred to the falling sand and the sand would fall under the combined effects of gravity and acceleration. An egg-timer would not be a good clock, a good measuring device, if acceleration was involved.

Another example of the inappropriate use of a measuring device might be the use of an old-fashioned pendulum clock on a moving carriage such as a train. A pendulum clock depends on the accurate movement of the swinging pendulum and when stationary is affected only by the gravitational field in which it operates. The swing of the pendulum would be seriously affected if forces other than gravity were added. For example, the forces generated by the motion of the train travelling uphill or downhill would affect the swing of the pendulum as would any bends or turns the train made. Pendulum clocks are good measuring devices when they are stationary but they cannot be used in an environment which involves movement or where there is no gravity. Egg-timers are excellent measuring devices when they are stationary or in constant motion but they cannot be used in an environment which involves acceleration or where there is no gravity. All measuring devices or clocks depend on their working environment.

An objection to the 'other clocks' argument by the supporters of Einstein's view would be that if the light clock is measuring time running slow so the other clocks **must** be doing that too but these non-light clocks fail to illustrate such time dilation in the way a light clock does. But why is the light clock superior to the other clocks? If all non-light clocks show no sign of time dilation what reason is there to prefer the evidence of the light clock? Do non-light clocks fail to demonstrate time dilation or does time dilation not exist and so no evidence can be found? The light clock thought experiment is no more than a thought experiment and there is no evidence that what can be 'thought up' in such an experiment does occur in the real world. The light clock thought experiment illustrates how time dilation **could** happen but it is not evidence that it **does** happen.

Time dilation has no basis in observation. The support for the view that all clocks must behave in the same way as a light clock comes not from experiment or observation but from theory or ideology. Time dilation is a hypothesis introduced by Einstein to resolve the contradiction in the views of Galileo and Maxwell. Galileo believed that speeds can be added together and Maxwell's believed that the speed of light was 'constant' and could not be added to or increased. As a result we have a dilemma: the views of Galileo and Maxwell are in conflict and it was this conflict which

Einstein believed was resolved by introducing space and time dilation.

Einstein's thought experiment reveals nothing more than the inappropriateness and quirkiness of using a light clock in this working environment and shows that this type of clock is inappropriate to such an environment. It does not show that time, or space, dilate. It is important to clarify these issues relating to measuring devices, clocks or rulers, and their working environment. After that, we shall return to explain in detail the Galileo/Maxwell problem and how Einstein believed time/space dilation resolved the problem.

Clocks and the Working Environment

A man with a watch knows what time it is.
A man with two watches is never sure.

A clock is a measuring device, a time measuring tool. Every measuring device must depend on the environment in which it works and that applies to light clocks just as it does to Swiss watches or atomic clocks. So, how does the working environment affect measurements of time?

There has been a debate for many years about whether travelling at speed would affect a very sensitive instrument such as an atomic clock. Part of the issue was settled in 1971 when an atomic clock was sent on a long journey at speed and the time which that clock recorded was compared with the time recorded by another atomic clock which did not travel at speed. This experiment confirmed that the two clocks recorded different time. The experiment made use of the fact that the Earth spins at about 1000 mph at the equator. Using the speed of the Earth's rotation one plane travelled at a net speed of 1500 mph and the other at a net speed of 500 mph, that is, the speed of rotation of the Earth plus or minus the speed of the plane depending on which way it flies. The two clocks in motion recorded time 'running slow' when compared with a clock stationary on the Earth's surface. 'Slow' in this context is not the special Einstein version of slow time described above. It is the everyday 'domestic' meaning indicating that two clocks show different times. The simplest explanation of this time difference in the two atomic clocks is that just as Swiss watches work in different ways in water and air, so atomic clocks work differently when moving at speed. We know that all clocks are affected by their working environment, even if the mechanism is not always immediately obvious. These atomic clock experiments do nothing to confirm Einstein's claim that time expands and contracts. They merely show that any mechanism is subject to the conditions of its working environment and atomic clocks are affected by the speed at which they travel while Swiss watches are affected by operating in water,

An atomic clock which operates at a great height provides a different measurement of time from a clock operating at a lower height. This is a result of the difference in the force of gravity at the two heights. Gravity is stronger nearer the centre of the Earth and exerts a greater influence on an atomic clock so provides a different record of time. Mechanical clocks are affected by operating in water or air and atomic clocks are affected by operating in strong or weak gravitational environments. The working environment of an atomic clock is so important that scientists are now looking at ways to establish an atomic clock in space where it will be free from the influence of gravity and the Earth's magnetic field.

In 2010, physicists at the National Institute of Standards and Technology (NIST), Colorado, carried out experiments with atomic clocks positioned at different heights, one about one foot higher than the other. The results confirmed what has been known from other experiments, that the clocks recorded different times as a result of this small height difference. Because of the height difference the clocks were at different distance from the centre of the Earth and the gravitational pull of the Earth. Despite the very small differences in height and so in the strength of gravity which affected the clocks a significant difference in their measurements was recoded.

These experiments with very precise atomic clocks show not that time runs faster or slower but that measurements made with measuring devices deliver different results depending on their working environment. This environment includes the speed at which they move and the force of gravity exerted on them, which is often determined by their height or distance from the centre of the Earth.

Every clock is a material object. The operation of any material instrument can be affected by environmental factors and speed is one such factor. Time measuring devices can be Swiss watches, sand in a glass, atomic or light clocks. Since they are all material objects they are all subject to the environment in which they are used. Human bodies are also a type of clock so it is possible that our bodies are affected by travelling at speed or living at high altitudes. But, even if this is the case, it has nothing to do with 'time' and the slowing of time. It confirms only that the human body can be understood to be a clock, a device for making time measurements, and, like all clocks, is affected by its operating environment.

In the next section it is argued that 'time' is a word we use to describe measurements of a particular type. Those measurements use measuring tools. These may be sand falling under the effect of the Earth's gravity or the swing of a pendulum or the 'beat' of a caesium atom. All tools must work in an environment and these time measurement tools also have working environments which affect their behaviour. It is not 'time' or 'space' which can dilate, expand or contract, it is our measurements which can vary because they rely on measuring tools which are affected by their working environment.

What is "Time"?

The debate about what time and space are has rumbled on for centuries. For a long time, scientists believed that space and time were objective, that they existed in their own right and the observer simply provided information about local occurrences of space and time. There were, it was believed, objective standards against which any local occurrences or measurements of time or space could be judged.

Einstein, however, believed that this was not the case. He believed that there was interaction between the observer and what was observed. Time and space, he believed, were not absolutes but were relative to the context in which they were observed. In Einstein's view, observations of time and space had a degree of subjectivity attached to them. In early school science experiments, children learn that how 'hot' an object feels depends on the hand doing the feeling. The hand which comes from a bath of cold water says 'hot' and the other hand coming from a bath of hot water says 'cold'. Einstein thought that our observations of space and time were relative to the context in which they were made and the observer who made the observations. However, Einstein never clearly explained *what* was being observed. It was argued above that a time measurement is not a measurement of **something** called time. Nor is a space measurement a measurement of **something** called space. Time and space are words in a special category. They are words used to describe measurements and the information provided by these measurements helps us to manage the world in which we live.

A simple example of measuring time is the old-fashioned egg timer with sand dropping from the top to the bottom. We can observe the sand falling from the top to the bottom of the timer. This information then allows us to boil our egg to perfection or to work out how long it takes me to walk from A to B. We are measuring time, making a time measurement, but this is not a measurement of **something** called time. This is a measurement which gives us time information. Consequently words such as dilate, expand or contract should not, and cannot, be applied to time. 'Time' is a word used to describe measurements of a special sort. Measurements can be wrong or distorted or partially right or wrong but they cannot expand or contract. They are also not measurements of entities which can expand or contract.

There is change in the world. Trees grow, leaves fall and the Sun 'moves round the sky'. We can measure these changes with measuring devices to provide information which we then use to control other events or changes in the world. We measure sand falling through a glass tube to tell us when an egg is boiled. We use the position of the Sun in the sky to tell us when it will be dark or when to meet a friend. These measurements are 'time' measurements but they are not measurements of something called 'time'. Time is the word we use to describe the gathering of information about changes in the physical world. To claim that such measurements are subject to dilation is a category error. It is the use of concepts from one category, objects, entities and matter, to describe what happens in another category, measurements and comparisons.

The problem is another example of Ryle's concept problem. Time is a measurement – a measurement of the changes in matter. Time cannot have properties such as expansion, dilation or bending. It is a special type of measurement which can be inaccurate or wrongly used but cannot expand, contract or be subject to any type of dilation.

The Bending of Space

Time dilation and space dilation are two sides of one coin. What has been said about the dilation of time also applies to space and we can now look at the arguments that space can dilate. It is sometimes not made clear by Einstein which meaning of 'space' is being used. Some examples and arguments seem to use space in its geometric sense of distance, the gap between material objects, and at other time it seems to be used to mean the 'fabric of space'. This ambiguity in possible meaning will be clarified as we examine the argument for the dilation of space. In a later section the 'fabric of space' meaning of 'space' is discussed at length. In this section the word 'space' is used in a geometric sense, the distance between objects.

In the traditional thought experiment, described above, which was intended to establish time dilation, observations of time were made from two frames of reference in motion relative to each other. Observations of space made from two different frames of reference, Einstein argued, also showed that space could contract or expand. An example of a frame in this context would be a train passing through a station. A passenger on the train sees from his window the station, platform and waiting passengers walking on the platform. We can call this the 'passenger frame'. The ticket collector on the platform sees the train moving and the passenger sitting in his seat drinking coffee. We can call this the 'platform frame'. Is the train really moving? Is the platform 'really moving'? No one frame of reference has priority over any other. There is no frame which can be used to say that the train is in motion and the platform is stationary. You may be sitting in a chair at home and that is your current reference frame. You believe that you are stationary, even though you know that the Earth is revolving on its axis and orbiting around the Sun and the Sun and the planet are part of a galaxy which is also in motion. There is no way we can establish who is in motion and who is stationary. There is no absolute frame of reference in the universe which we can use to judge who is in motion and who is stationary.

Einstein asked us to imagine a beam of light fired from one end of a moving carriage to the other. This beam is observed and measured by an observer in the carriage, the passenger frame. However, a second observer outside the carriage on the platform, the platform frame, would make a different observation and measurement. The observations are made from two frames. The observer in one frame believes he is stationary and sees the traveller on the train in motion. The traveller on the train believes he is stationary as the station moves past his window and his tea is at rest on the table.

The diagram below shows a light beam fired from A to B in a moving carriage and is observed by a traveller in the carriage and someone waiting on the station platform.

Diagram 8

Light in a Carriage

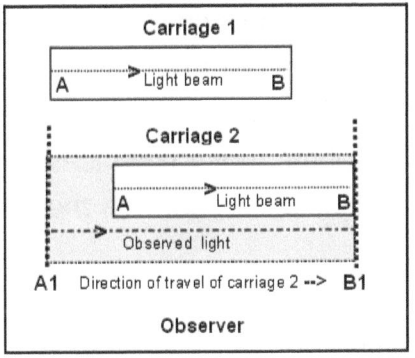

How is the speed of the light beam to be measured as the two observers see the beam from different frames and the speed appears to be different when viewed from these frames? The on-board observer sees his light beam travel from A to B but the observer stationary in relation to a moving train will see the beam travel A to B plus the distance the train travels which we shall call X. So the distance is A to B plus X, or A1 to B1.

The speed of light is constant and cannot change no matter which frame is used to observe it. From the passenger frame the light moves from A to B but from the platform frame the light beam travels from A1 to B1, which is from A to B plus the distance travelled by the carriage. If the measuring devices, the rulers, of the on-board and platform observers are to be in agreement then the distance A to B + X (A1 to B1) must be shortened to equal A to B. The faster the train travels the greater is the distance X and so the greater the contraction of space, or distance, must be. The faster the speed of an object the greater is the contraction of space. Since a material object has length and breadth those distances can also contract all material objects will shrink as they move at speed.

This is a variation on the time dilation argument. If a light clock in motion is observed and that clock ticks at the same rate as a stationary clock the light must travel a greater distance in the clock in motion than in the stationary clock. The answer is to shorten space so that the clock travels just the right amount of shortened (doctored) space to ping/chime in time with the stationary clock. Distance X in this example is the space equivalent of the distance C to C1 in the time illustration above (Diagram 6). Both time and space are subject to dilation when they are in motion, according to Einstein.

To the observer on the platform, the speed of the train should be added to the speed of light, but that cannot be done because the speed of light cannot be increased, it cannot be added to. This is the Galileo/Maxwell dilemma. The two speeds cannot be added but if we allow space to dilate we do not have to worry about adding speeds. We do not have to add the two speeds we simply say that the distance the light covers has contracted. The error in the thought experiment to establish that time and space dilate is that Einstein believed that the speed of light is constant, and this statement is open to misinterpretation and misunderstanding. To find an alternative to space and time dilation we need to move to the next step in the argument and examine what is meant by '*speed*' in the phrase 'the speed of light'. Although the speed of light is constant the speed of the carriage can be taken into account in measurements. The speed of light can be supplemented with other speeds to make measurements. The speed of the carriage does not increase the speed of light but both speeds can be brought into an overall measurement. It is not the speed of light which is at issue here but the meaning of the word 'speed'.

The Meaning of 'speed'

The established view that the 'speed of light' cannot be increased is mistaken or at least misleading. The speed of light is a crucial element in these discussions but the emphasis has been on the word 'light' in the phrase 'speed of light'. We need also to examine what the word 'speed' means. By providing a clearer explanation of what is meant by 'speed' we are able to remove the need for Einstein's argument that time and space can dilate. Many objects have built in mechanisms which limit their behaviour and speed. Take the example of a toy mechanical car which has been built with an internal mechanism which allows it to travel only at 20 mph. Its speed cannot be greater than 20 mph because it has its own internal, built-in limit.

Diagram 9

Speed of Mechanical Car

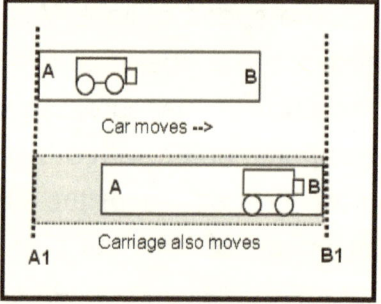

The speed of our imaginary toy car is 20 mph and it is limited by its own internal engine to that speed. But the car is located on a carriage which travels at 100 mph. The car moves from A to B on the carriage. The carriage also moves from A1 to B1 and the toy car, because that is how the experiment has been set up, arrives at B and B1 at the same instant. What is the speed of the car?

We can say that the speed of the car is really 20 mph because that is the limit imposed by its internal mechanism and that is its speed between A and B. But if the speed of the train carriage is 100 mph what is the speed of the car between A1 and B1? This is hardly an issue to be argued over. We have two possible definitions of speed of the toy car. First, the speed can be said to be time over the distance between A and B and the second speed of the toy car is measured over the distance between A1 and B1. Either meaning of 'speed' is acceptable so long as we understand which distances we are building into out calculation. Defining 'speed' involves a value judgement which depends on the purposes of the observer making the measurement. Different purposes justify different definitions and justify different elements being included in the calculation of speed.

If I am in the train carriage along with the toy car and I am waiting for the car to get to my end of the carriage to retrieve a message which is in the car, then I would prefer to judge the speed to be 20 mph, the speed of the toy car along the train carriage. However, if I am waiting at a station for the train and the car and the message then I might judge the car's speed to be 100+20 mph, the speed of the train carriage plus the speed of the toy car. The two answers to the question 'What is the car's speed?' are not mutually exclusive or contradictory. There is more than one answer to the question and which answer is preferred depends on a value judgement by the observer. 'The speed of ...' is a measurement which involves a value judgement. What we include in the calculation of 'speed' reflects the interests and motives of the observer.

Now, we can consider measuring the speed of light using an example similar to the toy car example. We have light travelling along a cable and the cable is coiled on the back of a moving carriage. Several kilometres of optical fibre cable are coiled and loaded on the back of a moving carriage.

Diagram 10

Light in a cable (1)

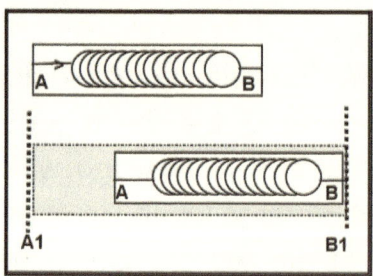

Several kilometres of cable are used only to provide, in an experimental situation, a long enough distance for the light to travel to make possible a comparison of the time the light takes from A to B and the time the carriage takes from A1 to B1. The length, however, is not crucial.

A beam of light is fired through the cable from the start, point A, to the end, point B. The light travels through the cable at the speed of light. In reality the speed is slightly slower than the speed of light as the speed is affected by the medium it travels through. However, the precise speed is not crucial in this example. Now we imagine that the carriage itself is in motion. The light travels both from A to B and from A1 to B1 so how do we measure the speed of light? Is it measured over the distance A to B or over the distance A1 to B1? The speed of the light in the cable is C and the speed of the light from A to B is C but the light has also travelled the extra distance represented by the distance A1 to B1 minus the distance A to B. This extra distance is the distance travelled by the carriage transporting the cable and light beam.

The solution to this problem is the same for light as it was for the toy car. The two possible answers about 'speed of light' are not inconsistent. They are compatible and depend on whether we choose to measure the light travelling from A to B, that is, from one end of the cable to the other, or to measure the light travelling from A1 to B1. If the light cable is attached to a light bulb and this is a signal when lit, but can only be seen by someone standing at B1, then if we are standing at B1 we need the light to travel both journeys, A to B and A1 to B1, for the bulb to light up and be seen at B1. To calculate when the bulb can be seen lit at B1 we need to know the speed of light in the cable and the speed of the carriage.

This gives us two meanings of 'the speed of light' which involves a calculation of both the speed of light in the cable and the speed of the carriage. The speed of light in the cable, in a container or carrier, is preserved at 186,000 mps while the container can be in motion at its own speed. Light, like the toy car, has a limited speed but if it is on a moving platform the speed of the platform can be used *in calculations* to supplement the speed of light itself.

Diagram 11

Light in Cable (2)

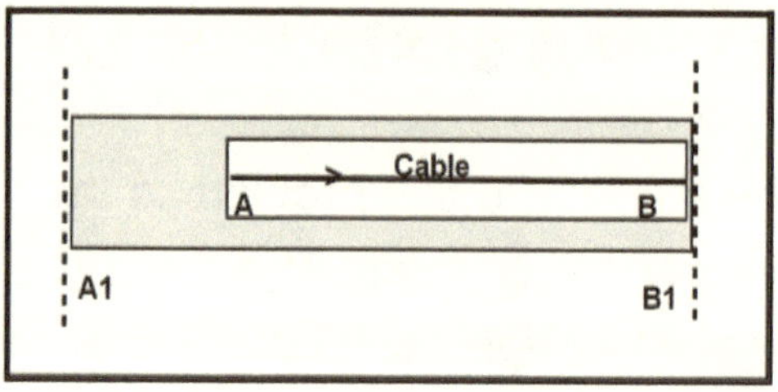

We can simplify this option and consider a train carriage with a cable stretched from one end to the other, from A to B. The logic involved here is identical to that which uses a coil of cable on the carriage. The light travelling in the cable from A to B as the train travels from A1 to B1 travels a distance greater than A to B. Because the train carriage is in motion the light also travels from A1 to B1. Even though the speed of the train, say 100 mph, is very small in comparison with the speed of light in the cable, the speed of light can be supplemented by the speed of the train in a calculation.

The next and final stage of this hypothesis is a light beam moving from A to B in the carriage but not in a cable. In this final example, the light beam in the carriage travels from A to B.

Diagram 12
Light beam in Carriage

The fact that a cable is not used does not change the event - the way the light beam behaves or its speed. The cable changes the medium through which the light travels but does not change the logic of the argument nor the speed or behaviour of the light beam.

As in the example of the coil of cable and the toy car on a carriage, the speed of light remains 'C' but the light can arrive at an end point of a journey at a speed of C but **supplemented** by the speed of the carriage. Of course, compared with the speed of light, the speed of the carriage in this example is almost negligible, but it is still an additional speed which can supplement the speed of light in the carriage when a calculation is made. There are two meanings of the phrase, 'the speed of light'. The first meaning of 'the speed of light' is uncontested: it is C or 186,000 mps, the speed of light itself, and cannot be increased. The second meaning of 'speed' will depend on what we want to include in the calculation of 'speed'. Such calculations can include the speed of light **supplemented** by the speed of a carriage or platform in motion. Supplementing the speed of light in this way resolves a long-standing problem in science, the Galileo/Maxwell dilemma, which is outlined in the next chapter. This problem and its solution were the main reasons for Einstein's introduction of the hypothesis of the dilation of time and space.

Chapter 3

The Galileo/Maxwell Dilemma

Einstein's model of space, time and gravity is better understood when put in the context of the views of three other eminent thinkers: Galileo, whose version of relativity was a precursor of Einstein's; the Scottish physicist James Clerk Maxwell, who provided an explanation of the nature of light, and Sir Karl Popper, who supported Einstein's theory of gravity on the grounds that it was testable and so scientific.

Before Einstein, Galileo had proposed a form of 'relativity', now referred to as Galilean Relativity, and this too made use of the concept of frames which was described above. Galileo asked how we should describe the events in the cabin of a ship which is travelling on a smooth sea. The sailor in the cabin sees that the lantern hangs from the ceiling of the cabin without swinging, his beer does not slosh out of his mug and when he drops his spoon it falls directly onto the table and not at an angle into his lap. He has no window to see the land passing but he does know the ship is moving because he has been up on deck before coming down to the cabin. He might, back in his cabin, expect his spoon when he drops it to be affected by the forward motion of the ship and to fall at an angle into his lap rather than directly down onto the table.

He drops the spoon and finds that it falls directly downwards onto the table.

Galileo argued that an object in motion continues in a straight line until it is affected by some force. The spoon starts to falls while the ship is in motion so it does not acquire any extra energy from the ship because the ship's motion is constant. Consequently the forward motion does not affect the spoon's fall. Acceleration, of course, would affect the fall of the spoon as an extra force would be applied as it was falling. The constant motion of the ship is there at the start of the fall of the spoon, is unchanged as the spoon falls, and so does not influence the fall. This is Galileo's Law of Inertia also known as Newton's First Law of Motion.

According to Galileo, a sailor in this situation where the beer does not spill, where the lantern hangs steadily from the ceiling and spoons fall directly to the table is entitled to assume that he is at rest, even though he knows that the cabin is in a ship which is in motion. The sailor is entitled to believe he is at rest though an observer on the shore would say he was in motion. The observer on the shore who sees the sailor in motion is standing on a planet rotating at some 1,000 miles per hour. So he is not at rest either though, like the sailor, he thinks he is at rest. Like the sailor, the observer's beer, as he stands on the shore, does not slosh out of the mug and spoons fall directly down. He too is entitled to think of himself at rest despite knowing about the motion of the planet.

There is no objective standard of being at rest or in motion. Consequently, any scientist carrying out an experiment in the cabin, or on the shore, is entitled to think that he is at rest and that the laws of physics remain the same in any laboratory which is at rest. The laws are the same in a laboratory on the shore or on the moving ship or on a modern train travelling at high speed. The laws of physics remain the same irrespective of the motion or speed of the laboratory where the experiment is conducted.

This Galilean view of the world was challenged by the work of the Scottish physicist James Clerk Maxwell who developed the equations which describe the propagation of electromagnetic waves such as light. These equations did not allow for the addition of velocities. The speed of light, he established, was constant and could not be added to. This did not fit with Galileo's view that speeds could be added together. The addition of speeds was central to Galileo's view. For example, according to Galileo, a boy on a ship travelling towards the shore at 5 mph fires a stone towards the shore at 10 mph. An observer on the shore measures the speed of the stone as 15 mph. The two velocities, that of the stone as it leaves the catapult and that of the ship, can be added and to the observer the speed of the stone is the two speeds added together. Such additions of speed could not be done if they involved Maxwell's equations describing electromagnetic radiation such as light.

According to Maxwell's equations, the speed of light was a constant. It was usually denoted by the letter 'C' indicating that it was a universal constant. A key feature of the speed of light is that it cannot be increased. In Galileo's view, speeds could be added together but, according to Maxwell's equations for light, the speed of light could not be increased. This destroyed the view inherent in Galilean relativity that the laws of physics were identical everywhere. According to Galileo, the observer should see the speed of light plus the speed of the ship but according to the Maxwell view he would observe the speed of light to be 'C' not 'C' plus the speed of the ship. The speed of the ship could not be added to the speed of light so the observer would observe an anomaly in an experiment carried out on a moving platform. The laws of physics would change depending on whether an experiment was carried out in a laboratory in motion or a laboratory judged to be at rest by the observer.

This conflict in the views of Galileo and Maxwell was a dilemma for scientists and the hypothesis that time and space could expand and contract was Einstein's attempt to reconcile these views of Galileo and James Clerk Maxwell. The hypothesis that time and space dilate, that they contract and expand, was not based on observation or experiment. The concept was introduced by Einstein to resolve this conflict of views. If space could dilate then it would not matter that the speed of the ship in motion and the speed of light could not be added together. The contraction of space means that if the ship is in motion the speed of this motion and the speed of light could be accounted for not by adding them but by showing that space contracts to just the right amount to make adding the speed of motion to the speed of light unnecessary.

"If a light clock in motion is observed and that clock ticks at the same rate as a stationary clock the light must travel a greater distance in the clock in motion than in the stationary clock. The answer is to shorten space so that the clock travels just the right amount of shortened (doctored) space to ping/chime in time with the stationary clock." (page 46 above)

Contraction of either time or space resolves this problem of adding the speed of a ship or any platform in motion to the speed of light. We have now arrived at the heart of the time/space dilation hypothesis. The views of both Maxwell and Galileo are valid and need not conflict if space or time can dilate. Space/time dilation is Einstein's resolution of this conflict of views of Maxwell and Galileo and this resolution is the reason he introduced the hypothesis of time/space dilation.

The hypothesis, however, is unnecessary if the speed of light can be supplemented in calculations and measurements with the speed of a carriage or platform in motion. The speed of light in a light clock remains at C whether the clock is stationary or in motion. If the clock is in motion it is not necessary to contract space or time to account for the extra distance covered by the light beam in the clock. The speed of the light beam in a clock in motion remains C but this speed can be supplemented with the speed of the platform carrying the light clock. Measurements involving the speed of light can be supplement with other speeds but the speed of light itself is not increased by those other speeds.

Frames, Observation and Events

Gravity is not affected by the motion of the ship, according to Galileo, because no extra energy is given to the force of gravity by the ship's movement. In a similar way, no extra energy is or can be added to a beam of light. The speed of light cannot be increased for two reasons. The mechanism which governs the propagation of light waves, as Maxwell's equations show, is a self-contained mechanism. In this respect, it is similar to the mechanism of the toy car – both are 'self-contained'. Neither mechanism can be 'cranked up' by adding energy from an outside source, such as the motion of a carriage. The consequence is that when an observer of a moving carriage, observing frame A from frame B, observes light travelling at C the speed of the carriage cannot be added because the speed of the carriage does not affect the internal mechanism of light, just as the speed of the carriage does not affect the speed of the toy car it carries. However, the speed of the carriage can affect the observation of light, and other waves such as sound, in another way.

Imagine a passenger on a train in motion who is playing a trumpet. The sound waves from a trumpet will not be affected if the trumpet is played in the carriage of the train and is heard by a passenger in that carriage. Both are in the same frame. However, if the trumpet is played on the moving train so that it can be heard by a listener on the station platform then this will create a Doppler effect. The speed of the sound waves will not be affected but the frequency will. The observer is not in the same frame as the trumpet and so observes the effect of motion on the sound. This Doppler effect also happens with light waves. The speed of light is not affected by the speed of the train but the wave length is affected. It was this Doppler effect on light from distant galaxies which Hubble observed and showed that galaxies were moving apart.

Frequency means that the sound waves arrive at different time intervals. Imagine a pop gun which fires ping-pong balls which travel at a speed of one yard per second. A boy standing one yard from the gun will receive a ball every second. If the boy starts to move away from the gun then he will receive a ball not every second but every second and a quarter and then every second and a half and so on depending on the speed at which he moves away. The further he moves from the gun the less frequently the balls arrive. The speed of the balls remains the same but because of the extra distance to travel their frequency of arrival decreases. Sound and light waves are affected in the same way if the person receiving them is moving away or towards the source of sound or light, or if the source is moving towards or away from the observer. We are used to observing the Doppler shift of sound in everyday life when we hear the siren of an emergency vehicle as it approaches us and then departs. The sound of the siren rises and falls and this is the result of the sound waves being contracted as the vehicle moves towards us and then the waves being expanded as the siren moves away.

The passenger in the carriage and so in the same frame as the trumpet player hears 'normal', not distorted, trumpet sounds. The observer standing on the platform as the train passes and so in a different frame notices a distortion of the sound, the Doppler effect. For both light and sound waves the frame of the observer establishes whether the frequency of the waves is affected. The speed, however, of the sound and light waves is not affected, only the frequency of the waves is affected. The frame determines whether the observer will hear the frequency shift. If the observer is in the frame he does not and if he is outside, in another frame, he does. The reason why the observer notices the Doppler effect is that the speed of the train is part of the observation of the sound waves in that frame. The sound waves do not change from frame A to frame B but the observation of waves, sound or light, is affected by factors such as the motion of the carriage and the frame from which the observations are made. 'Frame' is a concept which is important for observations. Frames do not affect events or what happens within the frame. They are relevant only to an observer's measurements and observation. Frames relate to observations only and not to the actual events being observed.

In the light clock thought experiment, which was supposed to establish that space and time can dilate, the two frames are not examples of two different sets of actual events. The frames establish only that **observations of events** from different frames will differ.

For the light clock which is not in motion we need only record or measure the speed of light in the clock. An observer in a second frame may want to supplement the speed of light in the clock with the speed of the platform carrying the clock. This supplementing of the speed of light with the speed of the platform does not increase the speed of light it merely builds the two speeds into a measurement.

Frames allow us to compare observations of events from one frame to another. Actual events, for example speeds, are not changed by being in frame A or frame B. The speed of light is not affected by the frame in which it is observed or from which it is observed. Einstein's thought experiment which claims that the speed of light in a clock in motion must increase with motion is simply wrong. The observation of the light clock by an observer in a different, stationary frame does not affect the speed of light. No actual events are affected by the observation from any frame. The option of supplementing speeds in measurements, but not adding them together, resolves the problem of the speed of light in a clock in relative motion. In doing so it removes the need to claim that time and space dilate. It resolves the Maxwell/Galileo dilemma which was Einstein's reason for introducing space/time dilation.

Science and Falsifiability

Einstein might have argued that non-light clocks must record dilated time because if all clocks are synchronised at the start with the light clock and the light clock runs slow when in motion then all other clocks must do so too. It is just difficult, the argument goes, to detect or prove this slow-running with any mechanical or non-light clock. This is the fatal flaw in the time dilation debate. Einstein's conclusions about time and space dilation are not based on observations or experimental results. They were introduced to resolve a theoretical problem in physics. That is why 'evidence' using non-light clocks is rejected. No evidence from observation or experiment, such as conventional clocks in motion, can prove that time dilation is wrong. Time dilation is a 'solution' to a dilemma posed by the opposition of the views of two scientists, Galileo and James Clerk Maxwell. If this is the case then Einstein's views are not scientific but are simply a reflection of his philosophical bias.

Sir Karl Popper was one of the most distinguished philosophers of science of the twentieth century and believed in Einstein because his theory that light was diverted through bent space was a testable theory. The Principe expedition tested the theory in practice and for Popper that meant that Einstein had put forward a scientific theory. Einstein's theory was testable and so was falsifiable and so was scientific. Popper approved of this and contrasted the Einstein approach to science with that of religion and pseudo sciences such as psychoanalysis and Marxism. The latter were impossible to falsify and were usually not exposed to the test of falsification by experiment or observation.

Popper accused Marxism of evading falsification. For example, Marx predicted that the first socialist revolution would take place in the most advanced industrial country, England. The first socialist revolution, however, occurred in the backward underdeveloped Russia. This turned the Marxist theory of historical materialism on its head. But Marxists refused to interpret the Russian revolution as a falsification of Marxist theory. They interpreted the Russian revolution to suit the requirements of their ideology. This made Marxism irrefutable and made it unfalsifiable.

Propositions which cannot be falsified are defended by adapting either theory or evidence or both to protect the proposition being criticised. Such propositions can never be proved wrong. They are unfalsifiable and, according to Popper, lose their right to be called scientific. It is ironical that when Einstein went beyond the Principe observations to propose a mechanism to explain this 'bending' of light he moved out of the realm of science. In taking this extra step from the observation on Principe to a model requiring time and space dilation, Einstein shifted his theory into the realm of the unfalsifiable and the unscientific.

Scientists have become used to understanding the world and gravity using a model which requires the dilation of time and space. In Einstein's words, quoted in the Preface,

"Concepts that have proven useful in ordering things easily achieve such an authority over us that we forget their earthly origins and accept them as unalterable givens."

Any contrary argument which denies time dilation, such as the use of non-light clocks, is rejected and observations are re-interpreted. For this reason time/space dilation, like Marxist theories, becomes irrefutable and so unscientific. If the argument in favour of space and time dilation is not falsifiable then it is not science. It is prejudice.Popper supported Einstein's explanation of gravity because it was testable and so falsifiable. It is ironic that this support should then be jeopardised by the model proposed by Einstein to explain these observations. The Principe observations appear to support the hypothesis that space is somehow distorted but the model proposed by Einstein, of time and space dilation and the bending of space, is wrong.

For the supporter of the time dilation hypothesis, conventional, mechanical clocks cannot be used to reject the light-clock argument that time can run slow or fast. The light clock is supreme because it allows Einstein to establish the slowing of time and this, in turn, resolves the Galileo/Maxwell paradox. The proposition that time and space dilate is not a scientific proposition, it is a prejudice disguised as science. The light clock helps illustrate how time dilation might work but it is not evidence that it really happens.

Once it is established that space cannot expand or contract and that there is no evidence from observation or experiment to support this view, the basis for the argument that gravity is the bending of space is removed. Without the possibility that space can bend, Einstein's claim that gravity is matter moving through bent space is an empty claim. It is, then, necessary to offer a new model of how gravity works which does not depend on bent space or time/space dilation. To do this, we need to understand what Quantum Mechanics has to tell us in this context.

Chapter 4

Space & Quantum Mechanics

Einstein did not always make clear what he meant by 'space', whether it was the geometry of space or the fabric of space. He also did not spell out the mechanism of how space was bent or how bent space interacted with matter to cause it to follow the path of bent space. Trains follow railway tracks and we understand the mechanics of how that works but Einstein did not describe how matter follows the track created by bent space – which we now know does not exist.

It is important to distinguish the 'fabric of space' from 'space' in a geometric sense. In the train carriage example outlined above, which is intended to show that space is bent, the geometric sense seems to be used. Often, it is simply not clear which meaning of 'space' is intended or being used. This section deals with 'space', in the sense of 'the fabric of space', which used to be called 'the void' or 'the vacuum'. This void is far from empty and is in fact seething with forces and activities. It is these forces, which are found in the fabric of space, which are vital to the explanation of what gravity is and explain the mechanism behind the observations made on Principe. Since Einstein wrote his paper on gravity, there have been many important developments in scientific thinking which help us understand how gravity works. The alternative model of gravity described below makes use of some options which quantum theory (QT) makes available.

Einstein believed his explanation of gravity as a result of the bending of space was adequate and he felt no need to look further. He may also have been inhibited from further enquiry because it would have taken him into the realm of quantum theory, which he never accepted as a valid scientific theory. Einstein's comment on quantum theory is usually translated from German as, "God does not play dice with the universe." Niels Bohr is said to have responded to this with, "Who are you to tell God what to do?". Indeed, Einstein's antipathy to QT and the whole idea of what he described as 'God playing dice with the universe' may have encouraged him to accept the simpler, almost 'mechanical' explanation of gravity which the bending of space model appeared to offer.

Unfortunately QT is difficult to understand. One of the most successful and respected scientists of the last century, Richard Feynman, who won a Nobel Prize for his work on quantum physics said:

'What I am going to tell you about is what we teach our physics students ... It is my task to convince you not to turn away because you don't understand it. You see my physics students don't understand it. ... That is because I don't understand it. Nobody does.' (Richard Feynman, Nobel Lecture, 1966).

Over forty years later, the situation has not changed: quantum physics remains as much a mystery now as it was then. To start to appreciate what quantum theory has to say we begin with what is known as 'The Two Slit Experiment', which illustrates the difference between the particle and wave explanations of energy and light. A simple illustration of how the particle model of force worked was set out above showing the particle as the carrier of the force. Now, we look at the wave model of force and how it differs from the particle model. To do this we must come to terms with several hypotheses which are counterintuitive and unpalatable to many.

For over two thousand years, since the time of Ancient Greece, there has been a debate about whether energy and light should be understood using a particle or wave model. The argument came into focus in the 1600s with the debate between Newton who believed that the particle model was the right model and the Dutch scientist Christiaan Huygens who supported a wave model. From then until the early 20th Century, most scientists agreed with Newton and preferred to understand light as made up of particles. In the 1920s the French scientist Prince Louis de Broglie persuaded many scientists that light was transmitted as a wave. At the start of the 20th century most scientists changed their views and started to think of light in terms of waves. It is a debate which has swung back and forth for hundreds of years and has still not ended. Physicists today, for most purposes, think of light as electromagnetic waves travelling at 186,000 mps. QT, however, introduced the possibility that both wave and particle views of energy are correct to some degree.

At the start of the 20th century, it was believed that particles were completely separate phenomena from waves but the double slit experiment showed that this view had to change: waves and particles were two ways of looking at one event. In our daily lives we expect to be able to identify objects and say where they are. In the world of sub-atomic particles, this is not the case. Heisenberg proposed a different way of looking at the world of sub-atomic particles. He proposed the Uncertainty Principle. This says that certain basic properties of an object, such as the position and momentum of a subatomic particle, cannot be measured simultaneously and with complete accuracy. Imagine a tennis ball flying across the court. The observer puts up an instrument which measures the momentum of the ball. However, this act of measurement interferes with the flight of the ball so further measurements cannot be made accurately. Any measurement will affect the object and will prevent or distort any later measurement. The Uncertainty Principles is the first hurdle to jump in understanding QT.

Some scientists argued that the measuring problem was simply the result of not very efficient measuring devices and that the devices could be improved to get rid of the problem. Others, however, believed that no matter how sophisticated the measuring equipment became the problem lay in the nature of the real world. Objects, such as subatomic particles, just did not have the qualities of exact location and momentum simultaneously. Improving the measuring devices would not change this. The ambiguity, it was argued, was built into reality, built into the nature of particles themselves and could not be removed by better measuring devices.

In addition, we have to take into account what is called the *complementarity principle*. This says that we must look at the world through wave spectacles or through particle spectacles. If we choose to gather information about the wave nature of particles this excludes gathering information about them as particles and vice versa. The Danish physicist Niels Bohr, who worked in Copenhagen, helped develop the standard explanation of what happens at the quantum level. This has become known as the Copenhagen interpretation and claims that we need to have two sets of rules. Nature is made up of fundamental dualities and the observer must select one of the dualities in preference to the other when making an observation. Light, he argued, can be interpreted as either waves or as particles and in any experiment the scientist must choose which model to work with. If you choose one model, for example the wave model, then that rules out using the other model, the particle model, in that experimental situation. Waves and particles are a reflection of the experimental preference of the scientist rather than a description of the way the world is.

As part of understanding the world of sub-atomic particles we have to take on board many counter-intuitive ideas. These ideas are an assault on our commonsense views of the world and the 'Two Slit Experiment' illustrates many of these peculiar aspects of the world of particles.

The Two Slit Experiment

The wave model provides us with a different view of reality from the particle model. When two particles collide they exchange energy rather like billiard balls. When two waves collide they combine and do not exchange energy like particles, instead the two waves merge to form new waves. To test the behaviour of waves a barrier with two slits is erected in front of a light source. These slits can both be open or one open and the other closed. The light is fired at the barrier, first in the form of particles and then in wave form, and the effect of the slits being open or closed is observed.

Diagram 13

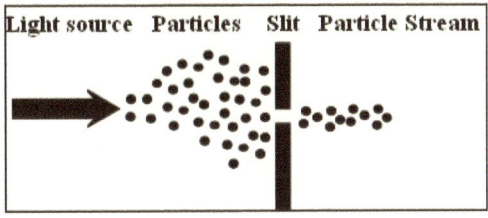

If light is fired at the barrier in particle form and with only one slit open it will behave in a bullet-like fashion and passes through the slit to form a pattern of individual marks like bullet marks on the receptor wall. But if light as a wave is fired at the barrier it passes through the slits and emerges in wave form, like circles or semi-circles as in the diagram below.

Diagram 14

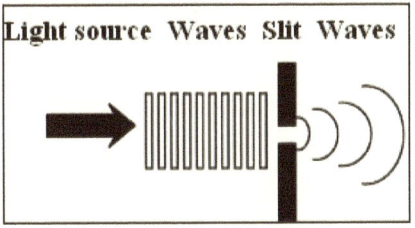

When two slits are open the waves which emerge from one slit interfere with the waves from the other slit, making an interference pattern which shows up on a receptor wall as a pattern of light and dark bands. Where the peak of one wave meets the peak of another wave they reinforce each other and pro-vide a bright band of light. Where the peak of one wave meets the trough of another they cancel each other out and produce a dark band on the screen.

Diagram 15

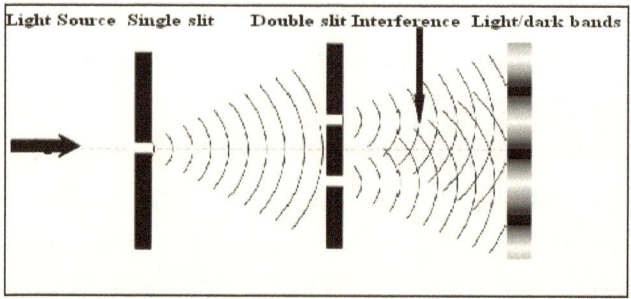

When the waves pass through the slits, they spread out and interfere with each other. According to the classical view of the particle model of light such interference would not happen and particles would exit from the slits as two streams without any interfer-ence between the streams. However, waves would interfere with each other as they passed through the slits.

In the two slit experiment, if only one photon, or particle of light, is fired at a time at the barrier with the two slits open we would expect the photon to pass through one slit or the other and create a bullet-like impression on the back wall according to which slit it passed through. The particle must go through one slit but cannot pass through both slits at the same time. If we wait until millions of photons have been fired, one at a time and pass through one slit or the other and start to build up a pattern on the second wall, we do not get two clusters opposite the two holes. Instead we get an interference pattern. It is as if each photon as it passes through one slit knows that the other slit is open and behaves as if it was a wave. Each individual photon places itself on the receptor wall in such a way that when enough have passed through, they build up an interference pattern when there cannot be any possibility of interference because each particle has passed through only one slit and on its own.

If we repeat the experiment but with only one slit open we get the result we would expect – a clustering of 'bullet marks' on the wall behind the slit through which the particles passed. Individual photons passing through one of the slits seem to *know* whether only one slit or two slits are open and if the second slit is open the particles behave as waves and form interference patterns and the typical dark and light banding appears on the final wall. If only one slit is open the photons behave like bullets.

This happens not just with photons and light but also in experiments carried out with particles such as electrons which, unlike photons, have mass. The double slit experiment is not just some sort of weird laboratory set-up which has no bearing on the real world. This odd behaviour of sub-atomic particles and waves is at the very heart of our understanding of the material world and how we manipulate that world. Quantum theory has found applications in the every-day applications such as television, computers and digital cameras. The Copenhagen interpretation explains this strange behaviour by a theory known as the 'collapse of the wave function'. According to the theory, what passes through the slits is not a material wave but a 'probability wave'. The particle does not have a definite location but has a greater or less probability of being at one place rather than another. Some locations will have a higher probability than others, such as the light areas in the band, and other areas, the dark areas, will have a low probability. In summary, an electron which is not being observed does not exist as a particle at all. It exists only as a wave-like property spread out over the area of probability where the particle will eventually be 'found'. In fact, the particle is not 'found', it does not exist yet and so is not there to be found. So how does this probability wave change to something real? The weird answer is that it is only when there is an observation of the waves made that they cease to be probable and become real – probability is converted to reality by the act of observation.

The American author and playwright Gertrude Stein lived most of her life in Paris. On revisiting her home town of Oakland, California, which she remembered with affection from her youth, she said,

'There is no 'there' there'.

The phrase records her disappointment, her failure to find the place of her youth. Her favourite places, the library, the park and meeting places, had disappeared and the town of her youth now existed more in her memory than in reality. Scientists could just as easily use this phrase to describe the experimental situation of quantum physics. Before the act of observation, there is no 'there' there. Before the act of observation reality is there, in some sense, but is not 'really' there until after the act of observation.

In an experimental situation, there are multiple possibilities and probabilities but no 'there', no actual entity, until an observation is made. After observation is made, the entity is brought into existence and is changed from being just a probability into a reality by the act of observing. Such a view of the world and reality verges on the mystical. It is a view we can understand from a playwright who is reminiscing about her past but more difficult to understand from a scientist. But that is what quantum physics requires.

A "particle" should be thought of as a wave function which has collapsed and after the collapse it is of such limited dimensions that it appears to be a conventional, classic particle, an entity with a specific length breath and thickness and with a spatial location which can be identified. However, even after the wave function has been collapsed, the particle which results from the collapse still has wave qualities and should be considered as a wave formation which is so limited that it appears to be and can be treated as a particle.

It is only when an observation is made of the particle/wave function that the probabilities collapse and the particle 'becomes' a reality. It is the act of observation by the observer which brings about this collapse of the wave function and brings into existence the actual particle. The two slit experiment introduces many counter-intuitive ideas including:

- 'ghostly' action at a distance
- the uncertainty and complementarity principles
- a particle is not just a particle but also a special form of a wave
- waves are not real waves but probability waves.

It may be difficult to accept many of these ideas but they are now widely accepted and lie at the heart of many advances in science and technology.

Einstein and Quantum Theory

At this point, it is worth repeating what Richard Feynman said about not understanding quantum theory. He said that although he did not understand quantum theory and it could not be understood, if scientists and engineers followed the instructions and the models provided by the theory they could achieve remarkable success in the real world not just the world of theory. This set of instruction for the quantum world has been dubbed the 'quantum cookbook. You do not understand what the cookbook means you just follow the recipes and you get results. The cookbook can be used but it describes a world which is 'incomprehensible'. It is the real world described by the theory, not the theory itself, which is incomprehensible and the source of puzzlement.

We have now arrived at the strange position where we have not just two models of sub-atomic particles, the particle and wave models, but where the waves are not real waves and the particles are not real particles. The waves are probability waves and these are collapsed by the act of observation and bring real particles into existence but those particles retain some qualities of waves.

Quantum theory provides a 'weird' and fascination world and Einstein never accepted this probabilistic views of reality. One reason for this reluctance was that one of his great successes had been to explain the photoelectric effect. This explanation of the way the photoelectric cell works was a major achievement for Einstein and his explanation required the particle model of light. This particle model relied on a Newtonian view of reality, a view in which particles had a physical existence, a location and momentum and this does not sit easily with a wave model with smeared out probabilities. It is not surprising that Einstein had a predilection for the particle view.

When light is shone onto a piece of metal the energy in the light causes a small current to flow through the metal. This is the photoelectric effect. The light passes its energy to the electrons in the atoms of the metal and this allows them to move around and creates the electrical current. However, not all colours have this effect on metals. If you shine a very bright red light on a metal it will not produce the photoelectric effect but a dim blue light will cause an electric current to be produced. This odd situation cannot be explained if light is thought of as a wave. Waves can carry as much energy as you want. Large waves carry a lot of energy and small waves carry very little but all waves carry some energy. If light is a wave then the brightness of the light determines the amount of energy it can produce. The brighter the light, the bigger the wave, the greater the energy and the greater the electrical current produced. Einstein realised that the answer to the problem was to treat light not as waves but as packets of energy. These packets were called photons and they behaved like particles. If we treat light as made up of photons then red light cannot dislodge electrons and cause a current to flow because individual photons of red light do not have enough energy. However, individual photons of blue light have more energy and so can dislodge electrons and cause a current to flow.

A bright red light emits lots of photons but each photon carries so little energy that it cannot move an electron and, no matter how many photons of red light there are, they will not have enough energy to create a current. If you hit a coconut with a peanut the coconut does not move because the peanut cannot carry enough energy. The coconut does not move if you bombard it with a thousand peanuts. Red light is made up of 'peanut' photons which do not carry enough energy to have any effect on the electrons in the metal. Einstein's work in this field showed that the photoelectric effect happened because light energy was only emitted or absorbed by electrons in discrete amounts, discrete packets of energy called 'quanta'. The 'amount' of energy in a packet could not be arbitrarily large or small. It had to come in fixed units or quanta. There is no smooth transition from one unit of energy to another. The units are discrete and there is a jump from one unit to another, not a gradual transition.

Einstein's view of the world was a classical Newtonian view in which science explained the causal links between chains of events which could be identified as discrete events. The Copenhagen interpretation of quantum mechanics says that while quantum mechanics can provide us with the rules for calculating probabilities, it cannot provide us with exact measurements of particles in the real world. Quanta are only a means of calculating probabilities but do not describe entities such as particles. Einstein could not accept such a probabilistic explanation as the best that physics could provide. He wanted to provide a complete causal, deterministic account of the natural world. Einstein's now famous phrase 'God does not play dice with the universe' summarises his antipathy to and suspicion of explanations of matter and its behaviour based on probabilities. He was very conservative in his support of classical Newtonian ideas and tried to raise several objections to quantum mechanics.

For scientists such as Heisenberg, an electron is a wave-like event which has the potential to change into a real event when observed. Einstein's interpretation of quantum mechanics was more statistical. He believed that an electron had a discrete position and momentum at all times during its travel. The wave function is only meaningful as a statistical description of the behaviour of a large number of particles but not of an individual particle. Individual particles, for Einstein, had the classical properties of position and momentum, even if it was difficult to discover those in an experimental situation. Einstein could never fully accept the probabilistic view of the world which quantum mechanics provided and struggled for many years to refute the basic assumptions of the quantum model, particularly action at a distance.

Action at a Distance and EPR

The two slit experiment introduces us to what appears to be 'action at a distance'. Action at a distance is when we have two objects which are separated in space and have no causal link but which behave as though they were linked to each other. In the Two Slit experiment the particle passing through open slit A seems to 'know' whether slit B is open or closed and changes its behaviour accordingly. Opening slit B affects the behaviour of the particle even though the particle does not pass through that slit. This suggests that the particle is part of a larger event which includes not just the slit A through which it passes but also slit B through which it does not pass but seems to 'be aware of' or have information about.

This idea of action at a distance was rejected by Einstein and he devised a thought experiment, the EPR experiment, named after the scientists who designed it, Einstein–Podolsky–Rosen. This experiment was proposed in a 1935 paper and was intended to show that quantum mechanics was not a complete physical theory. Quantum theory allowed for the possibility of action at a distance and Einstein believed that this alone showed that QT was not a valid scientific theory. With the advance in technology the EPR experiment was actually carried out by a French physicist, Alan Aspect, in the 1980s and the results proved that Einstein was wrong and quantum theory was right: action at a distance really did occur.

The Aspect experiment showed that particles separated by a very large distance and which could not 'communicate' or exchange information with each other still behaved as though they were part of a causal sequence but with no causal link between them.

Diagram 16
The EPR Experiment

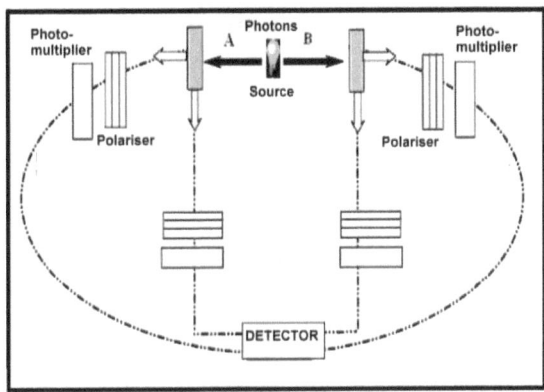

A simplified outline of the experiment is above. Photons are particles which are made up of vibrating waves oriented in different ways. For example, they can have different polarisation. If two photons are emitted from one source then they must, according to the rules of physics, have equal but opposite polarisations. We can visualise such polarisation as vibrating waves. The waves are either vertical or horizontal: they travel either in an upright, top to bottom fashion or in a horizontal, left to right fashion. The experiment uses polarisers which allow through only one type of wave and block the other. The polarisers are similar to ordinary sunglasses which allow through only certain types of light. The photomultipliers are just sensitive detectors which then amplify the signal and pass it on without altering the orientation. The gates are devices to direct the photons in one direction or the other.

When the two photons A and B with opposite orientations emerge from the central source they are diverted in different directions by the gates and taken towards a polariser. The gates through which the photons pass to get to the polarisers were wired so that they switched the way they directed the photons some 100 million times per second. This arrangement meant that while the photon was on its way towards the polarisers it would not 'know' which of the two polarisers it would go through and so it would not 'know' whether it would emerge as a vertical or horizontal wave. "Know" in this case just means "does not have access to information about...". As it passed through the system the photon would be too far away from its partner, the second photon, to be able to communicate with it. The distance between the two photons at the time of the switch by the gate/polariser would be so great that information from A about its polarisation could not be exchanged with B, even at the speed of light. The experiment showed that no matter how many times the orientation of the particles was switched they were able to 'communicate' this information to the partner photon and, when they arrived at the final detector, they both had opposite and so 'correct' orientation.

Despite the convoluted trip both particles had to take, despite each particle having its orientation switched many times and despite being unable to communicate their orientation to each other, the particles arrived at their destination with the opposite orientations as the laws of physics said they should. The Aspect experiment proved that quantum theory was right and that 'action at a distance' did in fact take place. Two particles can communicate instantaneously and change their orientation in response to a change in either particle. Quantum mechanics provided an accurate and complete explanation of what happened in the real world in this experiment. This explanation contradicted the view of classical Newtonian physics which Einstein preferred.

Quantum theory requires that the real world comes into being as a result of the collapse of probabilities and that those probabilities are spread out over a wide area. What happens in one part of that area of probability can effect, instantly what happens elsewhere and this is sometimes called 'action at a distance'. The causal link, which Einstein thought must exist, between two events, in this case the orientations of particle A and particle B, does not need to exist and indeed does not exist as this experiment showed.

The Casimir Force

The quantum view of the world opens up new possibilities for explaining the events which were witnessed in the Principe experiment and which were thought by Einstein to be a result of the 'bending of space'. Einstein believed that the 'bending of space' needed no further explanation. His antipathy to quantum mechanics probably also inhibited him looking in that direction for any further explanations of bent space. The 'fabric of space, however, as opposed to the 'geometry of space', when combined with QM provides new options which can explain the Principe observations.

Until some 100 years ago, scientists thought of 'space' or 'the void', much as the non-scientist did: an emptiness which did nothing and contributed nothing. This view, however, changed with the success of quantum theory which showed that the space is filled with 'virtual particles' which come into and go out of existence like bubbles appearing and disappearing on the surface of a liquid. Space is 'frothing' with particle energy which manifests itself from time to time in unusual ways. Above, it was described how a 'particle' could be thought of as a wave function which had collapsed and after the collapse was of such limited dimensions that it appeared to be a particle, an entity with a specific location which could be identified.

Even after the wave function had been collapsed by an observation the particle which is brought into existence still had wave qualities. It is these wave qualities which mean that particles in space have interactions with other particles and matter and form the basis of an unusual force. Such virtual particles are not just theoretical constructs. In the middle of the last century, a Dutch scientist, Hendrik Casimir, outlined an experiment to prove that virtual particles can exert a force and affect material objects in the real world.

The importance of the Casimir force in this discussion is that it is not a force in the conventional sense. It is a phenomenon produced by differences in particle energy in two areas of space and this difference produces results which *appear* to be a force, similar to magnetism. In the shipping business, a Casimir-like force has been known for hundreds of years. Waves or small ripples on the sea appear to cause ships to attract each other. The consequences of such attraction can be devastating. This strange force is capable of driving ships together and seriously damaging them. The force is created by a pressure difference between the ripples in the water between two ships and the more energetic ripples on the outside of the hulls of both ships. The pressure on the outside on the hulls, because it comes from waves in the open sea, is greater than the pressure between the hulls where there is limited scope for waves or ripples and limited energy is available. These excess forces on the outer sides of the hulls cause what appears to be an attractive force similar to the Casimir force. The pressure forces the ships together and the resistance from the water between the ships is not enough to compensate for the pressure from the water outside.

Some sixty years ago the Dutch scientist Hendrik Casimir claimed that the virtual particles postulated by quantum theory were not just theoretical constructs. They also had real effects in the world. Casimir suggested an experiment which showed that virtual particles could have effects in the real world. He asked what would happen if two metal plates were placed close together in a complete vacuum. In experiments with the Casimir force two plates are used and the quantum ripples of the electromagnetic field were shown to have a role similar to that of ripples on the sea. Evidence for the existence of this 'force' has been produced in the laboratory at several universities, notably St Andrews in Scotland. The focus of this Casimir research has been on nano-technology, the study of the very small.

Before quantum theory was generally accepted, physicists would have believed that the plates would just have remained stationary. Casimir believed that the forces generated by virtual particles would have a physical effect similar to the forces created in a conventional wave environment at sea. 'Space', 'outer space', 'the vacuum', 'the void', these are just a few of the words used to talk about the space between galaxies, planets and suns. This 'emptiness' is far from empty. The 'void' is alive with activity as virtual particles come into and fade out of existence. The vacuums holds an enormous amount of energy in the form of virtual particles and some scientists are considering how we might be able to tap and employ this vast resource just as we employ the winds to generate electricity.

It was not until 1997 that the technology became available to establish that such a force existed at the sub-atomic level and allowed scientists to measure it in the laboratory. These virtual particles are now well understood and have effects in the real world and practical applications. Much of the experimental work now being done deals with the force at the nano level but applications in the macro world are likely to follow as interest develops and research funds are made available.

Diagram 17
The Casimir Force - Wave Environment

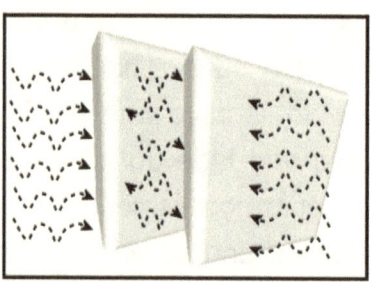

Casimir argued that virtual particles, which filled the so called 'empty space', would exert a small force which would cause two plates mounted in close proximity to move together. The force which drives the plates together is the result of an imbalance of energy created by an imbalance of energetic particles in the areas between the plates and outside the plates. This would be a force created out of nothing, out of the so-called empty vacuum, which would push the two plates together just as waves in the sea are thought to push ships together.

In laboratory experiments the amount of the force was incredibly small but nonetheless it was an actual force which resulted from virtual particles and had an effect on the real world, pushing together the metal plates in the experiment. Scientists are now certain that what was once called 'empty space' is in fact permeated by force waves which emerge from nowhere then disappear. Inserting two plates into this sea of energy creates three regions: one between the plates, where weaves energy is restricted, and two outside the plates, where energy is more abundant. The result is the creation of a pressure inequality which forces the two plates together as though attracted to each other by a force. The proof of a Casimir or Casimir-like force provides an alternative explanation of what Einstein thought was the bending of space. This alternative explanation identifies the expansion of the universe and the creation of areas of inequality of forces as the explanation of the force we call gravity.

Chapter 5

The Expanding Universe

A new model to explain how gravity works can be built on the hypothesis that the universe is expanding and that a Casimir-like force operates in the fabric of space. The expansion of the universe, of the fabric of space, provides the groundwork and foundation for a new model of gravity. It is now generally accepted by scientists that our universe is an expanding universe. This view was not always accepted. In Einstein's time, the steady-state theory was predominant and this theory described a universe which was static and whose size remained constant. The theory was supported by Ted Hoyle of Cambridge University. The

steady state view is that new matter was being con-
tinuously created as the universe expanded. Einstein's theory of relativity contradicted this accepted
cosmological theory and to avoid such a conflict with
the prevailing views of his time Einstein introduced
the 'cosmological constant' which mathematically removed the conflict. This allowed his theory of relativity
to fit with the steady-state theory in vogue at the time.
Essentially, Einstein cheated and rigged the mathematics and his equations to fit the accepted philosophy, the steady-state view of the world. Later, he accepted new evidence, mainly from the astronomer
Edwin Hubble, showing that the universe was expanding and not in a steady state. He then described
his cosmological constant as his greatest mistake and
rejected it. It was not Einstein's mathematics which
were at fault. The maths produced a view of the world
which was not fashionable and so Einstein decided to
'fix' his maths to agree with the accepted view. He
changed the maths to agree with a philosophical view
which he, at that time, believed was correct.

The steady-state view of the universe was over-
taken in 1929 when Hubble, working at the Carnegie
Observatory in California, measured what is called
the 'redshift' of some distant galaxies. It was his observations of this redshift which changed our view of
the universe. 'Redshift' is an example of the Doppler
shift which produces the distinctive sounds we are

used to of a police car or fire engine as it approaches and moves away. Exactly the same shift in wavelength is observed for light waves. Light waves from a distant galaxy are stretched out if that galaxy is moving away from us. This stretching out of the light waves shifts the light into the red part of the spectrum. If the light was moving towards us the wave length would be shortened and move into the blue part of the spectrum. So, the redshift provides astronomers with valuable information about the behaviour of the distant objects which are the source of the light and whether they are moving towards us or away from us. Hubble was also able to measure the distances involved and the only explanation of these observations was that the universe was expanding and distant galaxies were moving away from us.

Physicists concluded as a result of Hubble's observations that the separation between galaxies must have been smaller in the past and so by inference that at some time in the past they must have been bound together in one infinitely dense form from which the present universe was created by expansion. Everything in the universe had emerged from this dense and incredibly hot initial state in some sort of event or explosion, the start of the expansion of the universe, and this was called the Big Bang (BB). The Big Bang theory claims that our present universe expanded from an earlier universe which was not only much smaller, indeed infinitesimally small, but was also a very hot universe. As it expanded, the universe cooled as well as expanded. It was believed that the universe should be filled with the radiation which is the leftover of that early, very hot universe and the initial explosion or 'big bang'. This leftover radiation was called the 'cosmic microwave background' (CMB) or 'cosmic background radiation' (CBR). This left-over radiation, the afterglow of the Big Bang, does exist and was first identified by accident by two engineers working for Bell Telephone Laboratories in New Jersey, USA.

The two engineers were sent to try to sort out a problem with 'noise' in one of the Bell radio receivers. The engineers checked the equipment and thought the problem might be caused by the build-up of pigeon dropping in the large horn receiver. They cleaned out the bird mess and, some say, shot the pigeons to try to resolve the problem. However, after they had done all this they found that there was still a residual hiss or background noise which had no obvious source. The interference or 'noise' which was left after all the cleaning and adjustment had been finished was 100 times more intense than they expected and was spread with unusual consistency across the whole sky. They were certain that what they detected did not come from the Earth, the Sun or anywhere else in our galaxy. They both concluded that the noise must be coming from somewhere outside our own galaxy although they were not aware of any specific source which could account for this. By coincidence, scientists at nearby Princeton University were trying to devise an experiment to find the Cosmic Background Radiation and when they heard about the results of the Bell Lab work they realised that CBR had been found. CBR radiation is in the microwave sector of the electromagnetic spectrum and is not visible to the naked eye. The CBR fills the sky in every direction and its temperature is uniform to one part in a thousand. This uniformity as well as the universality of the radiation is persuasive evidence that this is the remnant of heat left over from the Big Bang. It would be hard to find a source of such radiation which had a local source and no one has managed to find an alternative explanation of this radiation.

The Big Bang preceded all galaxy formation. Galaxies did not exist at the start of the expansion of the universe and developed only later. Observations of galactic redshift established that galaxies were moving apart and it was this drifting apart of galaxies which resulted in the redshift which Hubble detected. Galaxies, which were formed many eons after BB, could not acquire the force of BB directly. In the moments immediately after BB, no galaxies existed and even the basic building blocks of galaxies, atoms and sub-atomic particles, did not exist. The energy of the

BB had to be transferred to these new galaxies and so affect their motion and their drift apart. The only intermediary, the only available candidate, for the transference of the energy of BB, is the fabric of space.

The fabric of space exerts traction on matter and it is this traction by the fabric of space which explains how galaxies, which were formed long after BB, are affected by the energy of BB and are still drawn apart in complex ways. The energy of BB is transferred via the fabric of space and the fabric of space exerts an influence on these galaxies. It is this influence which is described here as "traction". The fabric of space is responsible for the motion of galaxies as they move apart but Newton's Third Law tells us that the mutual forces of action and reaction between two bodies are equal and opposite: galaxies, suns, planets and moons, impose restraints on the fabric of space and the expansion of the universe.

Traction is not a one way street. A tractor which pulls a cart has its motion impeded by the resistance of the cart. The tractor pulls the cart and the cart pulls against, or restrains, the tractor. The situation with the fabric of space and celestial objects, suns, planets or moons, is similar.

Diagram 18
Spectrum of Traction

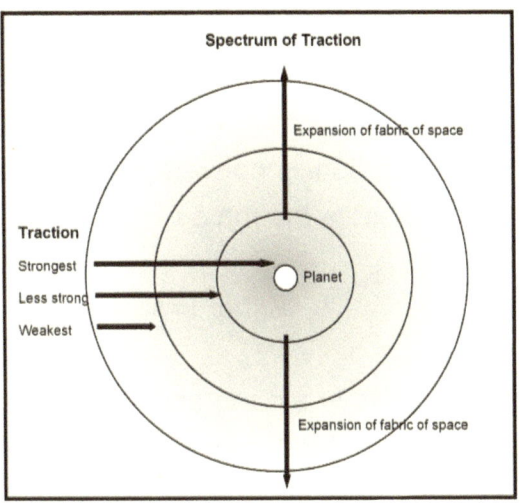

The diagram shows a two dimensional represen-
tation of the situation. The resistance by matter to the
expansion of the fabric of space does not inhibit the
expansion in general but it affects the expansion of
space locally. The area around planets and other celestial objects is changed by the presence locally of
matter. The 'grasp', or traction, which the fabric of
space has on matter, is at its strongest near matter.
This traction weakens as the distance from the material object increases. Planets, suns and apples inhibit
the expansion of space and cause a spectrum of traction around the object. The actual expansion of the
universe is in three dimensions and matter is embedded in this three-dimensional expansion. The traction

is also in three dimensions and this three dimensional area surrounding a planet, sun or other object is an area where the traction diminishes. The farther from the planet which is the source of the spectrum of traction, the less is the strength of the traction. The spectrum of traction affects material objects which are in motion. This motion is converted by the spectrum of traction and creates the force which we call gravity.

Matter in Motion

All matter is in motion and this motion may take one of two forms. First, planets, suns or galaxies are involved in multiple motions through space. The Earth spins on its own axis and around the Sun. Our solar system moves around the galaxy at 230 kilometres per second (km/s). In addition there is the motion of our galaxy. This can be broken down into the motion of the galaxy in the Local Group, a cluster of about twenty galaxies, and the motion of the Local Group in an even bigger group, the Local Super-cluster. When you add all these speeds together then you, as you sit in your chair, are travelling at about 900km/s.

Even travelling at some 900 km/s, the traction which surrounds the planet is a local phenomenon which moves along with the planet as it travels at a speed of 900km/s through the fabric of space. The situation is similar to the bow wave of a ship which moves with the ship as the ship travels through the water. The wave is not a fixed attachment to the ship. The wave dies and is re-created moment by moment as the ship moves through the water. Each wave has identical features even though the wave at time one is different from the wave at time two. The traction which surrounds a celestial object is, in a similar way, created anew from moment to moment as the object moves through the fabric of space. The spectrum of traction which surrounds a planet, like a ship's bow wave, does not vary or change as it dies and is recreated as the planet travels through the fabric of space.

In addition to this motion of objects moving through the fabric of space, all matter has a second form of motion which is the result of the atomic structure of matter. Even at the sub-atomic particle level matter is extended in space and is spread across areas of space and across a spectrum of traction. This form of motion, the motion of the atoms and particles which make up matter, is also subject to and converted by the traction around an object and becomes the force of gravity. It is the energy of matter in motion which is converted to the force of gravity.

Below is an illustration of the structure of an atom showing the atomic nucleus and the particles in motion around the nucleus.

Diagram 19
Particles in Motion

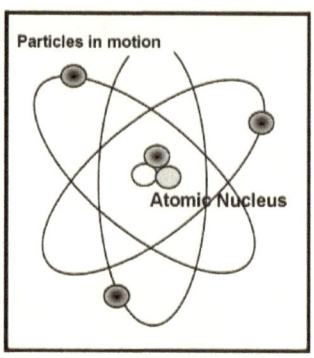

An atom and its sub-atomic particles, because they are extended in space, have some parts of their structure in areas of strong traction and other parts in areas of less traction. As a result, all matter, from sub-atomic particles to planets and apples, is spread across the spectrum of traction and moves to the area of greatest traction. The closer to the centre of the planet the greater is the strength of the traction. An apple hanging on a tree is a material object made up of atoms which, in turn, are made up of subatomic particles. These sub-atomic particles are in motion, in orbit, around the nucleus of the atom, and because they are in motion they, and the apple, are affected by the gradient of traction in space. The particles follow a path, created by the variation in traction, towards the centre of the planet. As a result of its atomic structure and the movement of the particles in the atom, the apple falls to the round when it loses the support of the branch.

Diagram 20

Gradient of Traction

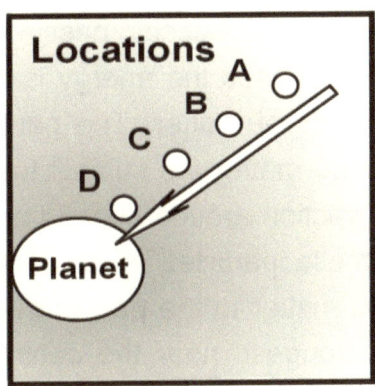

In the diagram above, matter is represented by the small white circles. Matter is shown following a gradient of traction which takes all matter to the point in the gradient where traction is strongest, the centre of a planet for example. Matter in this example could be any form of matter – an apple, an atom or a celestial body. All matter is spread across an area of traction and is driven by the energy of its motion towards the area of strongest traction. This gradient of traction surrounds all matter and this means that all matter has a gravitational 'attraction' for other matter. As the Moon attracts the water of the sea on Earth so the sea on Earth has a gravitational effect on the Moon.

Gravity is not the bending of space as Einstein claimed. Gravity is the result of inequality in the fabric of space. This inequality exploits the energy of motion of matter. This motion could be an object passing through an area of inequality, for example, a moon orbiting a sun or a spaceship passing near a celestial object. In most cases the energy is derived from the atomic structure of matter. The particles and atoms which make up matter are subject to the effect of the gradient of traction around planets and the energy of motion of these particles is converted into a force which drives matter to the part of the gradient where traction is strongest, near the centre of a planet or sun. This is gravity.

The book began with the observation that gravity is commonly understood to be a force. It concludes with the claim that gravity is, indeed, a force. As the universe expands, a Casimir-like inequality is produced in the fabric of space by the presence of matter which locally inhibits the expansion. This inequality results in the energy of matter in motion being converted to, and experienced by us as, the force of gravity.

Summary of Key Points

- Einstein claimed that time can run fast or slow and that space can expand or contract.
- There is no evidence from observation or experiment to support space/time (S/T) dilation.
- S/T dilation was needed for theoretical reasons, to resolve the Galileo/Maxwell dilemma.
- This dilemma can be resolved by showing that the speed of light can be supplemented.
- Einstein claimed that gravity was not a force but the result of the bending of space.
- If space cannot bend then Einstein's explanation of gravity is wrong.
- Gravity is the result of the unevenness in different areas of the fabric of space.
- This unevenness is partly the result of the expansion of the universe.
- The expansion is inhibited locally by matter and produces a gradient of traction.
- All matter is in motion as a result of its movement in space and/or its atomic structure.
- Energy from either type of motion is converted to a force by the traction of space.
- This is the force of gravity.

Appendix A

Spacetime

In addition to the concepts of space and time, Einstein often referred to 'spacetime' because he believed that space and time formed a four-dimensional continuum. We normally think of matter in our world as having three dimensions – length, breadth and thickness. However, we really need to add a fourth dimension, the dimension of time, to make our description of the world complete.

We can better understand this concept of spacetime by imagining a situation where we want to meet a friend and have to specify the meeting place. Suppose we agree to meet the friend at a certain place and we describe how to get there. We need to provide three pieces of information. We can specify the location on a map by specifying a point where a north/south line intersects with an east/west line: meet at the corner of Main Street and North Road. However, we may need more than the two dimensions which a flat map offers. If the specified location is a tower block with many floors we needs to specify on which floor to meet and so a third dimension is brought in. Even that, however, is still not enough. If we follow the three dimension direction we may not meet our friend unless we specify when we are to meet. There is little point in my arriving at the specified location on Monday if my friend does not arrive until Tuesday. If we do not add the fourth dimension of time to fix our meeting then the chances are the friend will be 'lost in time', that is, the friend will arrive when I am not there. So we need a specification with four dimensions.

It is this four dimension specification which is usually known as 'spacetime'. There is nothing mysterious about spacetime. It is essentially the three dimensions of space with the dimension of time added. Spacetime has no characteristics which give it special status or which mean we have to treat it is some way different from the other dimensions when they are treated individually. There is no mystery attached to it and no need to treat it in any special way.

Appendix B

Mathematical Models

Many textbooks and reference sources use mathematical models to explain spacetime and the distortion of spacetime. Mathematics is very flexible and malleable. If you ask a mathematician to describe a three dimensional world using a mathematical model, he can do that. If you ask a mathematician to extend this model to describe a four dimensional world he can also do that. These models of multi-dimensional worlds can be further extended to describe six, or seven, or eleven or twenty-one dimensional worlds. Mathematicians can add dimensions to their work simply by plugging them in. From the point of view of the mathematician it is not a difficult exercise.

Physicists like James Clerk Maxwell made enormous contributions to the development of science and mathematical physics. The danger, however, of a mathematical model of the real world is that it is a double edged sword. Mathematical models are easy to change and manipulate but they may mislead us as well as inform us about the nature of the world. Mathematics works because it extrapolates from established knowledge or information gathered from the real world. Using mathematics to create models of the world, to tell us what the real world is like is a risky venture.

Einstein initially produced a mathematical model of the universe which showed it was expanding. But the accepted view at the time was that it was static. As a result, he adjusted his mathematical model to fit the accepted view of the world as it then was. He added a 'cosmological constant' which made his equations produce an outcome showing a static universe. He adjusted his mathematical model not for mathematical reasons but simply to fit the accepted view of how the world was. Later, he described this adjustment, the 'cosmological constant', as one of his worst mistakes. He removed the cosmological constant and went back to a mathematical model which showed an expanding universe. This was 'mathematical 'sleight of hand', a fix, on the part of Einstein. He had chosen to prefer the philosophical model of a non-expanding, static universe and support this with mathematics. When this model proved to be wrong he dropped it and the associated mathematics and provided new maths to support his new philosophical view of an expanding universe. It was not a mathematical error but a philosophical error which he later supported by mathematics.

The reasons he preferred one mathematical model to the other were not mathematical. The reasons were both philosophical and practical. They were the results of observations of the real world by observers such as Hubble. It was these practical observations which supported a new philosophical view of the universe and showed it was expanding. The mathematical modelling then followed later to reflect this new view. Mathematics is good at reflecting models of the world which are established by other methods, such as experiments, observation or philosophy, but it has questionable validity in trying to create such models on its own. Einstein said of mathematical propositions and their relationship to the real world, *'they are not certain; and as far as they are certain, they do not refer to reality'*. By that he meant that some mathematical propositions can be certain because they are tautologies: 1+1=2 because the meaning of 2 is 1+1. But maths which purports to describe the real world has to be treated with caution and some suspicion.

The arguments above have dealt with space and time as discrete concepts and the arguments are not affected by the use of the unified concept of 'spacetime'. The fact that some mathematicians describe a spacetime which is distorted can only be treated as an interesting feature of mathematical modelling rather than a description of the real world. It is to the real world we need to turn to examine what space and time are and what meaning we can give to the claim that space and time dilate.

Even the enormously successful mathematical equations of Maxwell, which explain the behaviour of light, need to be understood in a wider, philosophical context. The meaning of the word 'speed', when used in the phrase 'the speed of light', may include not just the speed of light itself but other related measurements which can be built into a calculation of the speed of light. When we try to apply maths, even the very successful maths of Maxwell, to the real world we must take into account how such mathematics relate to associated concepts such as 'speed' as illustrated in the arguments above.

Appendix C

Muons and Time Dilation

It has been argued that the behaviour of muons is a proof of time dilation. Some fifty years ago, it was thought that experimental evidence had been discovered which proved that time dilation actually took place. The evidence came from the detection of subatomic particles called muons. These particles start their life at the outer edge of our atmosphere as incoming cosmic rays come into contact with the air. They are highly unstable particles with a half-life of 1.5 millionth of a second or 1.5 microseconds. A half-life of 1.5 microseconds means that if you have 100 muons at a given time then 1.5 microseconds later there will be only 50 and 1.5 microseconds after that you will have 25 and so on.

In 1941, a muon detector was placed at the top of Mount Washington in the USA, about 6000 feet above sea level. This counted some about 600 muons per hour. Given the half-life of muons observers expected only about 35 per hour to survive when they reached sea level. When the detector was brought down to sea level instead of the calculated 35 per hour it detected 400 per hour. The reason so many survived, it is claimed, was that in the muon frame of reference less time had passed because of the high speed at which they travelled. Their travel to sea level should have taken only 6 microseconds but, because of the purported time dilation factor, calculated to be about 9, the muon clock registers its speed divided by 9. This gives a time of 0.67 microseconds for the journey from 6000 feet to sea level. In this period of time only about one quarter would decay. In the time of 0.67 microseconds, which the muons experience in their frame of reference, Mount Washington is only 670 feet high and not 6000. Time and/or space are contracted for the muon because of the speed at which it travels.

This is not proof of time dilation. It is one possible explanation among several possible explanations and does not amount, on its own, to a proof of time dilation. If time dilation was a proven theory then this might be an instance of such dilation. As other possible explanations can be provided this should be considered as one piece of evidence only for time dilation but not as proof. The behaviour of muons is a puzzle but it should not be resolved by reaching for an off-the-shelf explanation which is not proven experimentally. It may be evidence to support time dilation but it does not prove it. Time dilation has been left unquestioned for over a hundred years and it has been invoked too easily as a convenient tool to explain certain events.

All matter is a form of time measuring device or clock and all such clocks are dependent on their working environment, Muons are also time measuring devices and they too are affected by their working environment. The behaviour of muons no more establishes time dilation than the behaviour of atomic clocks travelling at speed establishes time dilation. The behaviour of the muon, like the behaviour of an atomic clock travelling at speed or la Swiss watch operating in water, is no more than an indication that all clocks are affected by their working environments. Exactly which aspect of the working environment of the muon affects its decay has yet to be established. Such behaviour, however, is certainly not proof of, nor evidence for, time dilation.

Addendum

As this book was being prepared for publication, it was announced (BBC News, 23 September 2011) that researchers at CERN have been able to increase the speed of neutrinos to a speed greater than that of light. The findings are subject to confirmation and the researchers have asked other laboratories to try to replicate their work.